Power Systems 4.0

Current Trends and Future Outlook

Mutegi Mbae and Nnamdi Nwulu

CRC Press
Taylor & Francis Group
Boca Raton London New York

CRC Press is an imprint of the
Taylor & Francis Group, an **informa** business

First edition published 2025
by CRC Press
2385 NW Executive Center Drive, Suite 320, Boca Raton FL 33431

and by CRC Press
4 Park Square, Milton Park, Abingdon, Oxon, OX14 4RN

CRC Press is an imprint of Taylor & Francis Group, LLC

ISBN: 978-1-032-52789-5 (hbk)
ISBN: 978-1-032-66528-3 (pbk)
ISBN: 978-1-032-66529-0 (ebk)

DOI: 10.1201/9781032665290

Typeset in Times
by Apex CoVantage, LLC

Contents

About the Authors

Mutegi Mbae is a consulting electrical engineer with over 15 years of experience mainly in the power distribution and retail segments at Kenya Power and several other power utilities in Africa and Europe. Besides that, he is passionate about research, innovations, universal access to energy, zero emissions, power system optimization technologies, and global best practices in the energy sector.

He has authored a number of peer-reviewed, high-quality journal papers and presented a number of conference papers. Mutegi has also authored several book chapters in the area of power systems.

He holds a doctorate degree in electrical and electronics engineering (power systems) from the University of Johannesburg in South Africa and BSc and MSc degrees in the same area of study. He is an adjunct in the area of power systems at the Jomo Kenyatta University of Agriculture and Technology in Kenya.

Nnamdi Nwulu is a professor in the Department of Electrical and Electronic Engineering Science at the University of Johannesburg. He is a versatile professional with a strong background in research, education, and engineering. Holding both a BSc and an MSc in electrical and electronics engineering, and a PhD in electrical engineering, he possesses a comprehensive academic foundation. His expertise lies in the application of digital technologies, mathematical optimization techniques, and machine learning algorithms within the domains of food, energy, and water systems. With a keen interest in advancing these sectors, he has actively pursued research opportunities to drive innovation and enhance efficiency.

As a passionate educationist, he is committed to sharing knowledge and fostering learning experiences. He brings a multidisciplinary approach, combining technical proficiency with a dedication to impactful solutions.

He has authored several books, inventions, and over 200 publications, winning him the 2021 NRF Research Excellence Award for Early Career/Emerging Researchers in engineering/technology. He was also a 2020/2021 TW Kambule-NSTF-South32 Award finalist in the emerging researcher category and won the 2020 UJ Vice-Chancellors' most promising young researcher award. He is a member of the prestigious South African Young Academy of Science (SAYAS).

He is a professional engineer registered with the Engineering Council of South Africa (ECSA), a senior member of the Institute of Electrical and Electronics Engineers (SMIEEE), a senior member of the South African Institute of Electrical

Engineers (SMSAIEE), and a Y-rated researcher by the National Research Foundation of South Africa.

Prof. Nwulu is the editor-in-chief of the *Journal of Digital Food, Energy & Water Systems* (JD-FEWS), associate editor of the IET Renewable Power Generation (IET-RPG) and the *African Journal of Science, Technology, Innovation and Development* (AJSTID).

Preface

In its January 2022 Electricity Market Report, the International Energy Agency (IEA) contends that the global electricity demand surged by more than 6% during the year 2021. This was the largest increase since the 2010 global financial meltdown. As a matter of fact, the importance of a robust, agile, stable, and well-controlled power system cannot be over-emphasized.

This book presents the journey of the advancement of power systems, evolution of power transmission and distribution systems, system modeling, system control and optimization, upsurge in the uptake of renewables, and finally the advent of Industry 4.0 and its role in the management of power systems. The book also takes a look at the place of the emerging Society 5.0 in the management of power systems.

The book sequentially looks at the ongoing grid modernization and grid optimization and control, all the way to the aspects of deeper penetration of renewables and the aspects of the Fourth Industrial Revolution in power systems. The book seeks to tackle the emerging trends, utility business models, new technological advances, and deeper power system penetration of renewable energy, and thus is best suited for readers who want to get deeper into the area of power systems, both practice and research.

1 Fundamentals of Power Systems

1.1 THE EVOLUTION OF POWER SYSTEMS

Electricity refers to the movement of electrons in an electric circuit. Electric energy is generated by any of the three general methods below:

- Conversion of mechanical energy to electrical energy. This occurs in hydro plants; wind power plants; ocean waves; geothermal generators; coal-, gas-, or diesel-fired thermal generation machines; biomass plants; and nuclear power plants. The common principle here is the use of mechanical energy in running water, wind, and steam to rotate turbines inside an electric field to generate electric current as governed by Faraday's first and second laws of electromagnetic induction.
- The chemical reaction or electrochemistry that occurs mainly in batteries and fuel cells entails the conversion of chemical energy into electricity.
- Solid-state conversion exemplified in solar plants is the conversion of light to electrical energy by photoelectric cells [1, 2].

This is illustrated in Figure 1.1.

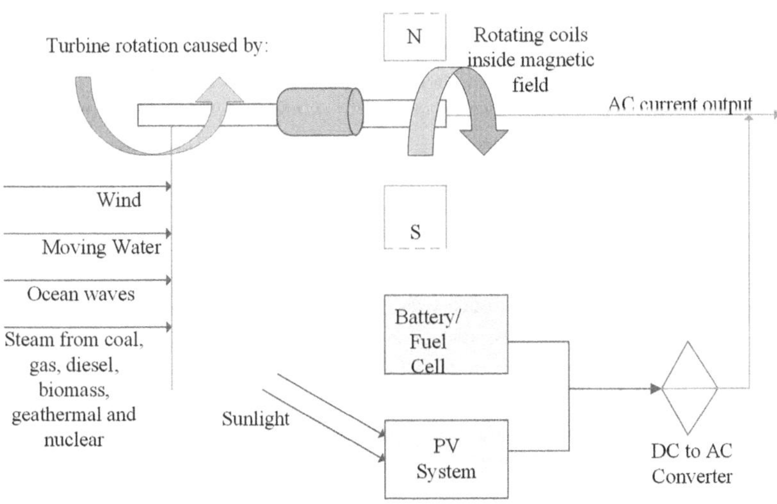

FIGURE 1.1 Methods of Power Generation.

DOI: 10.1201/9781032665290-1

The century between 1750 and 1850 saw the advent of many groundbreaking discoveries that laid the foundation for the development of power systems as we know them today. The discoveries were mainly in the area of electromagnetic conversion and power generation by researchers such as Alessandro Volta, Carl Gauss, Joseph Henry, Michael Faraday, and Nikola Tesla, among others.

In 1831, Michael Faraday and Joseph Henry documented their groundbreaking experiment, wherein they showed that an electric current is produced when a conductor revolves around a magnetic field. This has been the working principle of power generators and, indeed, the bulk of power generation many years later. In 1839, Edmond Becquerel developed the first photovoltaic solar generation using an electrolyte cell.

In 1870, Zenobe Gramme designed and developed a dynamo that produced a steady DC source, which was the inspiration behind the idea of using electric power to light homes. Power systems, as we know them today, have come a long way over the years. The first water wheel-powered station was built in England in 1881 to power 250-V arc lamps. This was closely followed by a DC steam-powered generating station in the United States created by Thomas Edison in 1882. In 1887, Deptford opened the first 10,000-hp AC power generating station in United Kingdom from which power was transmitted to London consumers at 10 KV.

The first long-distance power transmission system was built in 1891, which supplied 200 KVA energy over a 15-KV AC transmission line over a distance of 175 KM from a 210-KVA, 95-V, 40-Hz, 150-RPM synchronous hydro power plant in Germany. In 1896, the long-running practice of placing power generators away from load centers was started with the construction of the Niagara Falls hydroelectric power plant. The generated power was transmitted to Buffalo in New York, 35 km away. This was the first demonstration of the superiority of AC systems over DC systems in terms of carrying and distributing power over a long distance. This was done by George Westinghouse, largely using technology that was pioneered by Nikola Tesla.

In the formative years, as stated above, electricity was developed as a source of lighting for homes and streets. In fact, Thomas Edison patented the light bulb in January 1880. Electric companies initially billed customers based on the number of light bulbs in use and not on the actual KWH usage. Many power companies are still called lighting companies instead of electric utility companies. Many households had no internal wiring, so the utilities did the same for their customers besides selling power to them. With no other electric appliances, lighting was the primary use of electricity for at least a decade.

During the formative years, there was a vicious war (aptly named the war of currents) between proponents of the use of AC or DC for power transmission and distribution. The former was promoted by George Westinghouse, and the latter group was led by Thomas Edison. The invention of the AC transformer in 1882 by Lucien Gaulard and John Dixon catapulted the AC promoters to the driving wheel for many years by allowing power to be generated, transmitted, and distributed over long distances. However, all households still had to use Edison's light bulb. The first major 11-KV AC line (still in use by many power distribution utilities) was done from Niagara Falls to supply Buffalo, a distance of 32 KM, in the United States in 1895.

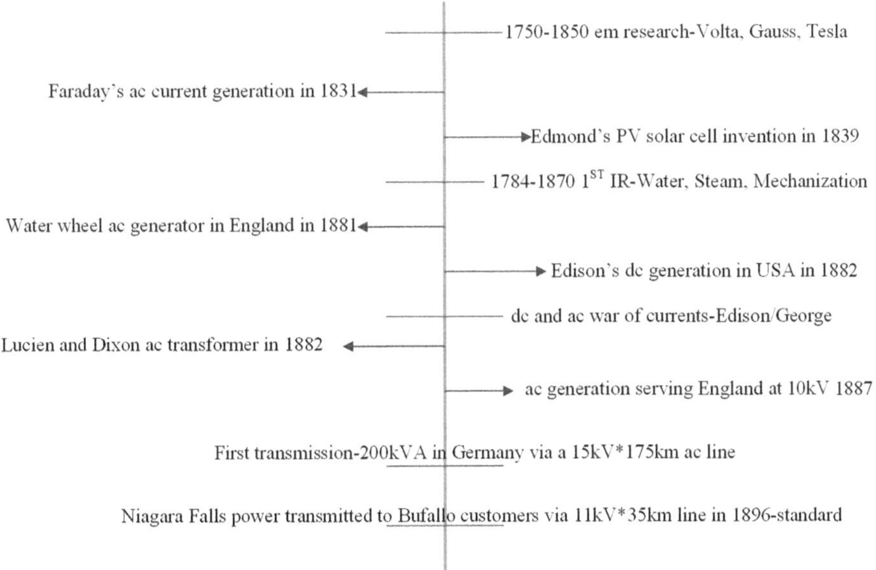

FIGURE 1.2 Early Evolution of Power Systems.

Westinghouse's concept of AC power transmission and distribution has remained largely unchanged more than a century later, except for efficiency improvements and a number of innovations and inventions [2–5].

The early evolution of power systems is summarized in Figure 1.2.

Edison's empire gradually gave birth to General Electric, and Westinghouse's business gave birth to the ABB Group as we know it today. This was because both entrepreneurs started to become uncomfortable with the increasing government regulation in the electric retail segment as power became a necessity rather than a mark of prestige. Equipment manufacturing had less control and the kind of engineering culture and profit margins that both craved [4].

The choice of frequency for AC power generation, transmission, and distribution has been a trade-off between various competing system needs. In the initial years, the frequencies ranged from 25 Hz to 133 Hz. For generators, the lower the frequency, the lower the number of magnetic poles required. The same is desirable for the transmission network because line reactance is directly proportional to frequency. The goal is to minimize technical energy losses on the line. On the other hand, higher frequencies are desirable for most loads. A compromise standard of 60 Hz for the Americas and 50 Hz for most of Europe was adopted.

Years later, the advent of high voltage direct current (HVDC) system in 1936 brought back the use of DC systems for commercial use in bulk energy transfer over long distances, more so with the increase in connectivity between diverse grids, countries, regions, and power pools. The first commercial use of HVDC was in 1954 in Sweden to interconnect the mainland to the island of Gotland using a 100-KV*96-KM HVDC submarine cable.

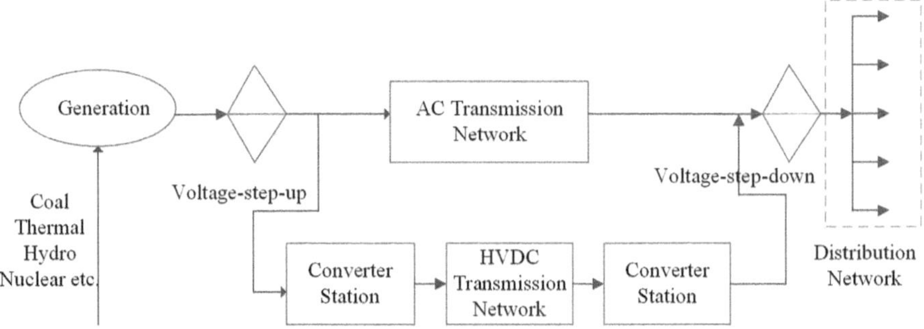

FIGURE 1.3 Convectional AC and HVDC Power System.

The development of grid codes and standards has evolved over the years. This started with the establishment of the Central Electricity Board (CEB) by an act of parliament in 1926 in Britain. The CEB established the first ever 132 KV, 50 Hz synchronized AC national grid for serving England. Power systems have slowly evolved to convectional electric grids that have been in existence for more than 100 years. For the longest, the grid has revolved around the concept of economies of scale, where energy is generated in big stations, transmitted in bulk, and finally stepped down for distribution to end users, as illustrated in Figure 1.3.

AC systems still have an edge because of voltage step up or step down, thus are still predominant systems at the generation and distribution stages brought about by the invention of transformers. This is facing serious disruption with the deployment of smart grids, distributed generators, aggregators, virtual power plants, and virtual transmission lines, among other forces of power system disruption as will be seen in detail later in the book. Many industries such as transport, hotel, and manufacturing, among others, have faced disruption brought about by the emergence of communication technologies and the fast growing Fourth Industrial Revolution. The power sector will not be an exception to this.

Years later in the 1990s, privatization was brought back to spur competition, attract capital, and improve on the quality of service. The idea of creating competition and efficiency in the generation and retail segments was to help lower the cost of power and provide an array of other customer services, e.g., prosumers and data services. Each retailer would receive the best energy prices as determined by market forces from various power generating companies. The jury is still out on the development of a robust and competitive retail segment in many countries and utilities vis-à-vis the age-old vertical power utilities, as will be seen later in this book.

The convection power system has been overtaken by events. This grid has become old, outdated, and unable to handle the new realities of power generation, transmission, and control. Integrating variable renewable technologies into the grid, bidirectional flow of energy, energy storage systems, decentralized control, prosumers, and the emergence of energy democracy in the wake of power system Utilities 2.0 is a big challenge for the convectional grid.

The above historical perspective brings out the realization that power systems have remained largely unchanged with not much new inventions for well over a hundred years. It is only with the recent advances of computer technologies that we have witnessed increased innovations and a change in the way power systems are operated, e.g., load flow for better system management and planning, and remote system operation. However, the core power system equipment and components remain largely the same, with most improvements centered around equipment operational efficiency. The importance of power system modeling, virtual simulations, sensors, power electronic devices, SCADA, satellites, and wireless technologies – all converging around the Internet of Things (IoT) – has continued to grow exponentially [6–10].

The term IoT was coined by General Electric in the year 2012 to describe multiple devices and equipment connected via communication technologies to collect, collate, analyze, and exchange data for real-time decision-making. IoT has heralded a new era of unlimited possibilities for power systems. This ranges from big data analytics, artificial intelligence, machine learning, better grid control and stability, to deeper penetration of variable renewables, among other numerous benefits.

The term smart grid was first used in 2005 in the Institute of Electrical and Electronics Engineers Power and Energy Society's (IEEE PES) *Power and Energy Magazine*. It was later adopted into law in the 2007 Energy Independence and Security Act. The letter and spirit of the law envisioned a versatile grid that ensures a great level of system security, quality of power supply, system reliability, environment friendliness, sustainability, efficiency, openness and transparency, interactivity, and availability [4].

The smart grid largely consists of generating stations with a deep penetration of renewable energy and storage, robust sensing and monitoring capability, and control to accommodate two-way sharing and the use of information rooted in energy democracy and a strong and secure 5G communication system. This may just herald the much-awaited large-scale wireless transmission of electricity in the coming years.

Interestingly, smart grid has brought back Edison's concept of small generators serving given areas of a neighborhood. The resurgence of distributed generation, rise in prosumer phenomena, the emergence of virtual power plants, micro-grids, and other emerging technologies have had power systems come full circle [8].

The First Industrial Revolution was driven by water and steam, resulting in the mechanization of agriculture. The Second Industrial Revolution was accelerated by the invention of electric power. Digital technologies played a pivotal role in the Third Industrial Revolution, resulting in the automation of many sectors in the wake of mass production.

The Fourth Industrial Revolution (4IR), or Industry 4.0, is primarily powered by the confluence of technologies, with no clear lines between the physical, digital, and biological spheres. This has resulted in promising areas of big data, quantum computing, cloud computing, intelligent cobotics, augmented reality, nanotechnology, bioinformatics, genetic engineering, genome editing, artificial intelligence, additive manufacturing, autonomous vehicles, and neurotechnology. 4IR will add value to power systems in the critical areas of digitization, decarbonization, democratization, and decentralization. It will ultimately deepen automated grid control and coordination, optimal generation, explosion of renewables, real-time system diagnosis, better system visualization, and predictive maintenance, among other benefits, resulting in Energy 4.0 or Utilities 2.0.

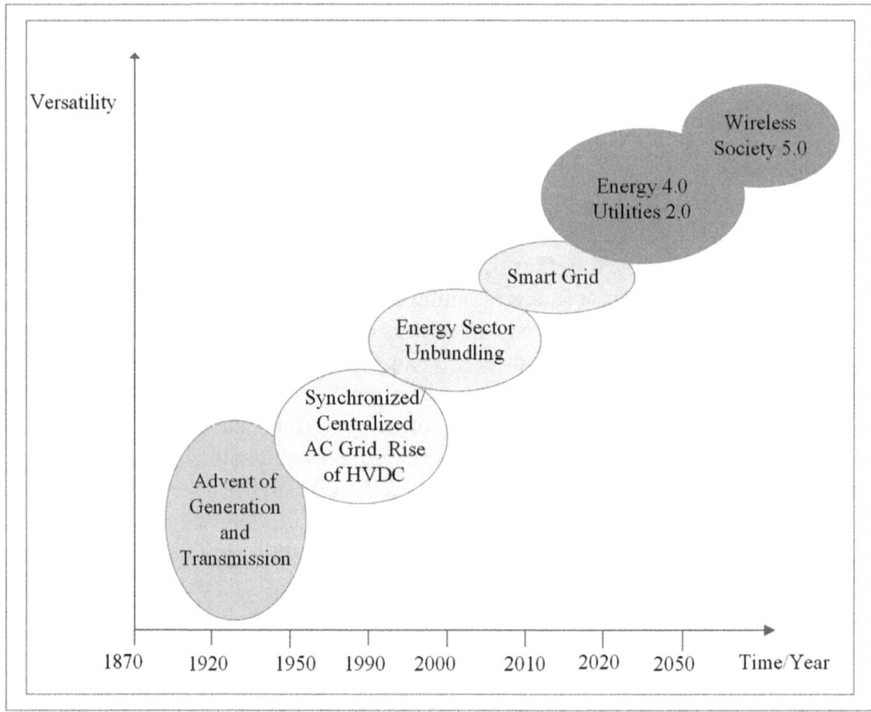

FIGURE 1.4 Evolution of Power Systems.

Wireless energy transmission and distribution should be the next frontier in the evolution of the power system. Figure 1.4 summarizes the evolution of power systems over time.

1.2 POWER SYSTEM FUNDAMENTALS AND CONCEPTS

1.2.1 Circuits and Network Analysis

A network refers to the interconnection of two or more devices such as a voltage source, resistor, inductor, and capacitor.

A branch is a simple path that has one device, e.g., a capacitor or a resistor.

When a network has at least one closed loop, it becomes a circuit, as seen in Figure 1.5.

A mesh is a standalone loop, meaning it does not enclose any other loops.

An active network contains at least one active device in terms of a voltage or current source. As such, a passive network contains none of these two [11–14].

A node is the point of connection for two or more active or passive devices in a power system, as illustrated in Figure 1.6.

In a linear circuit, the parameters are constant, meaning they are not a function of changes in voltage or current. On the other hand, the parameters of nonlinear circuits change with voltage and current.

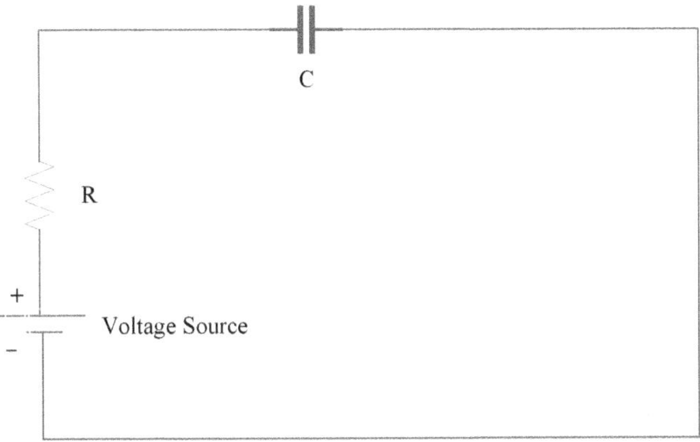

FIGURE 1.5 Simple Circuit.

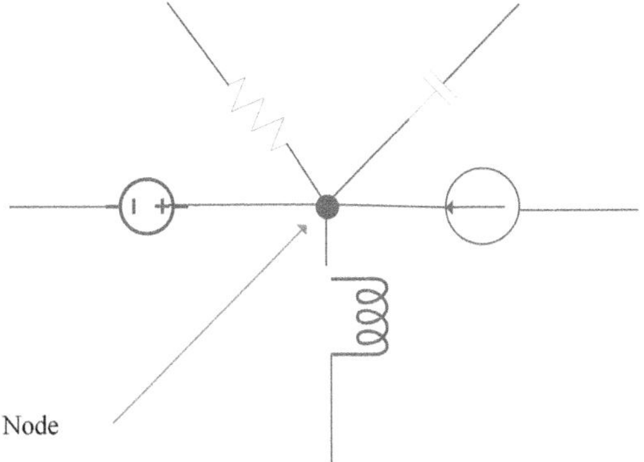

FIGURE 1.6 Node in a Power System.

1.2.2 Circuit Theorems for Linear Circuits

Kirchoff's current law states that the sum of all currents flowing into any node in a circuit is equal to the total current leaving the said node, as seen in Figure 1.7 [14].

Mathematically, the above is described as follows:

$$I_1 + I_2 = I3 \tag{1.1}$$

The Kirchoff voltage law states that the summation of voltage drops around a closed loop in a given circuit is equal to zero.

Thevenin's theorem states that a linear circuit with two terminals can be reduced to a voltage source (Thevenin voltage) in series with a resistor (Thevenin resistance), as shown in Figure 1.8.

V_{th} is the open-circuit voltage at the terminals, and R_{th} is the equivalent resistance at the terminals. During the transformation process, all independent sources are turned off.

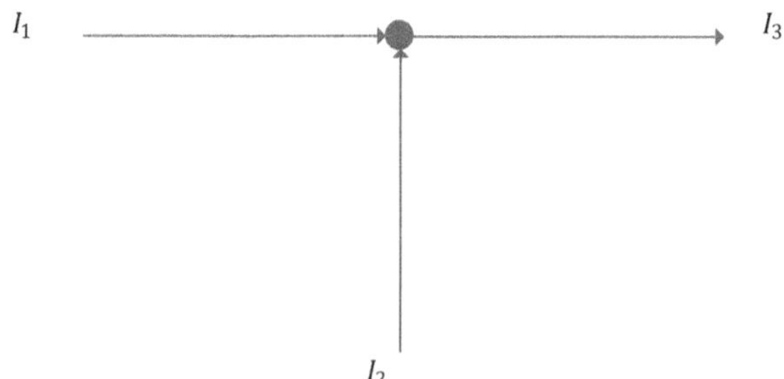

FIGURE 1.7 Illustration of Kirchoff's Current Law.

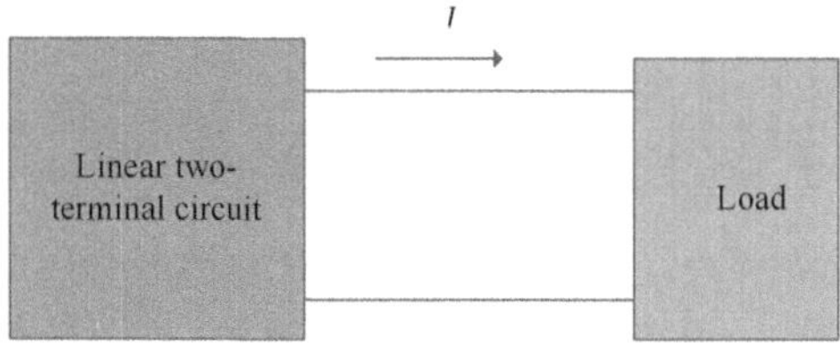

FIGURE 1.8 Thevenin Transformation of Electric Circuits.

Norton's theorem states that a linear two-terminal circuit can be reduced to an equivalent circuit consisting of a current source (short-circuit current) I_N in parallel with a resistor R_N with all independent sources turned off, as seen in Figure 1.9.

Millman's theorem states that any number of parallel voltage sources can be reduced to one using the transformation in Figure 1.10.

$$E_{eq} = \frac{\pm \frac{E_1}{R_1} \pm \frac{E_2}{R_2} \pm \frac{E_3}{R_3} \pm \ldots \pm \frac{E_N}{R_N}}{\frac{1}{R_1} + \frac{1}{R_2} + \frac{1}{R_3} + \ldots + \frac{1}{R_N}} \tag{1.2}$$

If N is the number of nodes in a given circuit, then the N-1 nodal equations are formulated to fully describe the circuit.

Given $L = M$ = number of loops in a given circuit, B = number of branches, N = number of nodes in the circuit, then $L = M$ = independent loop or mesh equations for the circuit to fully describe the circuit formulated by setting the sum of the voltage drops around each of the loops or meshes $L = M = B - N + 1$ equal to zero.

1.2.3 Basic Units of Measurement in Power Systems

A system is a structured group of building blocks of interconnected elements that work in synchronism according to a set of boundary conditions and rules to achieve

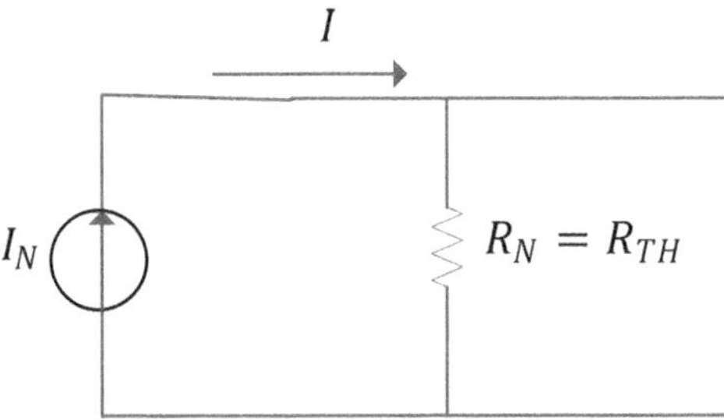

FIGURE 1.9 Norton's Transformation of an Electric Circuit.

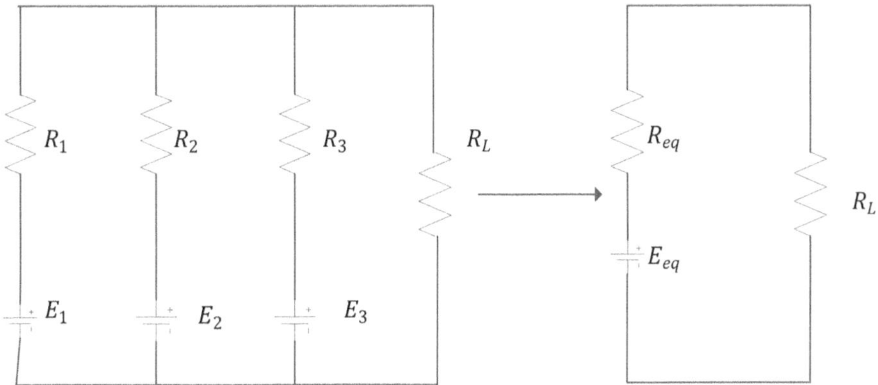

FIGURE 1.10 Illustration of Millman's Theorem.

a given purpose. In the study of power systems, it is important to break it down to critical parts to assist in understanding of its operation. Each part or building block of the power system has to play its role in order for the entire system to function. The system is built from the basic fundamentals upward to one of the most complex man-made engineering systems [15–18].

Electric current is the flow of charge (electrons) through a conductor in an electric circuit. It is measured in ampere. One ampere is the current flowing in two conductors, one meter apart and placed in a vacuum to produce a force of $2*10^{-7}$ newton per meter between the two conductors. The current can be constant (direct current) or alternating (alternating current). This is illustrated in Figure 1.11.

AC current and voltage represent the mathematical functions of the trigonometric sine and cosine functions. This can be represented by three parameters, namely, amplitude, frequency, and phase.

We use the root mean square (rms) or effective values of the AC voltage or current waves to measure the actual current or voltage supplied. This is computed as

$$V_{RMS} = \frac{V_{max}}{\sqrt{2}}. \quad (1.3)$$

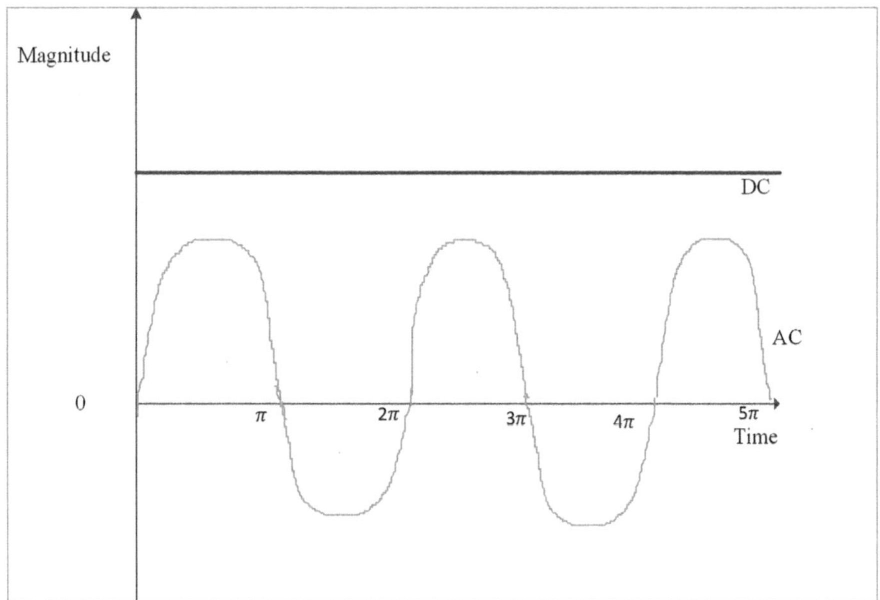

FIGURE 1.11 AC and DC Currents.

The concept of rms is employed when a variable averages to zero, but its effect is not zero. Squaring makes the negative values disappear, whereas the square root undoes the squaring operation.

The rms values are what the power utility companies are mandated to supply to their customers, as shown in Figure 1.12.

The mathematical formula is

$$v(t) = V_p \cos(2\pi f + \theta),$$

where V_p = the peak voltage,
f = the system frequency,
t = time (semi),
θ = phase of the sinusoid (degrees).

For an AC current or voltage, the average value is the mean of all instantaneous values for one alternation (zero to peak and back to zero again). They are DC values. As such, average voltage for one alteration = 0.636 (peak voltage).

Two sinusoidal AC waves are said to be in phase if they are in step with each other. This implies that both waves must cross their maximum and minimum points simultaneously. Otherwise, one wave lags or leads the other wave by any number of degrees.

If two sinusoidal waves are in step, we describe them as being in phase; else, one wave leads or lags the other.

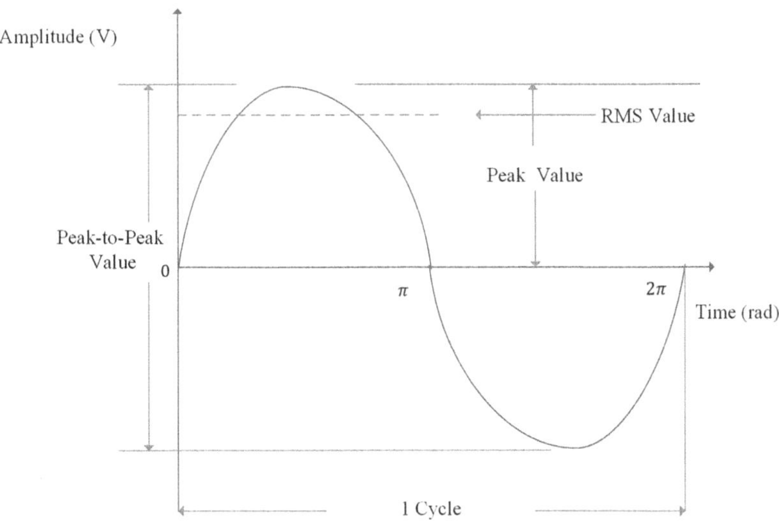

FIGURE 1.12 AC Waveform.

Current is one of the seven base units of measurement of the International System of Units (SI) from which other units of measurement are derived. The others are length, mass, time, temperature, luminous intensity, and amount of substance.

Frequency is the number of complete cycles or revolutions obtained by alternating current or voltage in one second; e.g., a sine wave that does two revolutions in one second is said to have a frequency of 2 Hz.

$$\text{Angular frequency (rad/s)} = 2\pi f = \frac{2\pi}{T}\text{; } T = \text{period (sec)}$$

Period T refers to the time it takes for an AC wave to complete one revolution or cycle. It is the reciprocal of frequency, i.e., period = 1/frequency.

Voltage is the potential energy of a power system. This means that it does nothing, but it has the potential to do work. This potential is what pushes/pulls electrons through an electric circuit. It is a push or force that is normally between two points, with one point acting as the reference point. The voltage can be constant (direct voltage) or alternating (alternating voltage).

Separately, voltage and current do not work. However, when both are combined to produce power, work is done. Power (measured in watts) refers to the rate at which work is being done, resulting in heat dissipation, lighting, and resultant mechanical energy output.

Electrical power is the rate of use of energy, calculated as the product of voltage and current. It is converted to useful output (work) by transferring or transforming energy from one form to another, causing wasted energy. It is utilized or dissipated as heat, light, or mechanical energy per given time frame, measured in watts. One watt is the amount of power that causes the production of energy at a rate of one joule per second, i.e., one watt = one joule per second.

TABLE 1.1
Units of Measurement in Power Systems

Quantity	Unit	Symbol
Electric capacitance	Farad	F
Electric charge	Coulomb	C
Electric conductance	Siemens	S
Electric resistance	Ohm	Ω
Illuminance	Lux	LX
Inductance	Henry	H
Luminous flux	Lumen	LM
Magnetic flux	Weber	Wb
Magnetic field strength	Ampere per meter	A/m
Magnetic flux density	Tesla	T
Energy (work)	Joule	J
Force	Newton	N
Heat	Joule	J
Length	Meter	M
Mass	Kilogram	kg
Pressure	Pascal	Pa
Resistivity	Ohm-meter	Ωm
Temperature	Kelvin	K
Torque	Newton-meter	N-M
Angle	Radian	Rad

Electrical energy (measured in watt-seconds) refers to the total power delivered over a given period of time. It is the capacity to do work and is a product of power and time. It is the sum of power delivered over a period of time and is measured in joule (watt-sec). One joule is the work done when one newton is applied to a point and displaced by one meter in the direction of force, i.e., one joule = one newton meter. It can be in many forms, including potential energy, kinetic energy, electrical energy, mechanical energy, chemical energy, and heat energy.

Table 1.1 shows a number of other derived units of measurement in power systems.

1.2.4 Impedance in Power Systems

Impedance refers to the opposition to flow of electric current in a power system. The most basic opposition is called resistance. A purely resistive circuit offers the same amount of opposition to flow of current in both AC and DC power systems. In DC power systems,

$$\text{Resistance}(R) = \frac{\text{Voltage}(V)}{\text{Current}(I)} \tag{1.4}$$

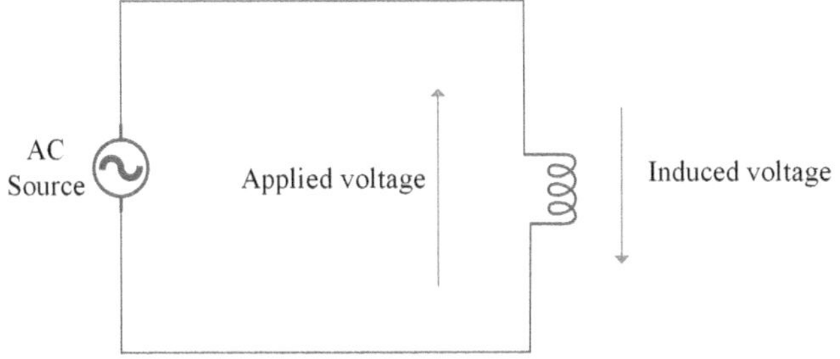

FIGURE 1.13 Occurrence of Inductive Reactance.

$$\text{Power}\left(P\right)=\text{Voltage}\left(V\right)*\text{Current}\left(I\right) \tag{1.5}$$

The above formulas show that in DC circuits, voltage and current are directly proportional. On the other hand, resistance and current are inversely proportional. In AC power systems, purely resistive circuits are rare [17].

A current flowing through an inductive coil produces a magnetic field around the coil. The amount of magnetic field is directly proportional to the amount of current flowing. This magnetic field produces a voltage that opposes the flow of current through the said coil—an inductor opposes any change in the current flow. This current flow opposition of an inductor is called inductive reactance. Inductive reactance occasioned by the counter-electromotive force (emf) is produced when a magnetic field develops around a current-carrying coil of a conductor, as illustrated in Figure 1.13.

Inductance is computed as

$$\text{Inductance } L\left(H\right)=\frac{N^2\mu A}{l}, \tag{1.6}$$

where N = number of turns of the wire,
A = cross-sectional area,
L = length,
μ = material permeability.

An inductor opposes the change in current, and on the other hand, a capacitor opposes the change in voltage.

A given circuit has 1 henry inductance if a current changing at a rate of 1 A/s produces an induced counter-emf of 1 V.

The voltage across an inductor is directly proportional to the rate of change in current flowing through it, which is given as follows:

$$v_L=L\frac{di}{dt}=2\pi fLI_p\sin\left(2\pi ft+90^o\right). \tag{1.7}$$

For a given inductive circuit carrying AC power, the inductance opposes the current flow. This opposition is called inductive reactance (X_L). It is computed as

$$X_L=2\pi fL, \tag{1.8}$$

where, f is the frequency of the AC source (Hz) and
L is the inductance of the circuit (H).

Capacitors are used to perform certain functions on electrical circuits because of their ability to charge and discharge in a circuit carrying AC current. For a given capacitor, capacitance C (measured in farad F) is a function of the size of conductive plates, their separation, and the type of insulating material used. One farad is the capacitance produced when a 1 V potential causes a 1 coulomb electrical discharge to accumulate on a given capacitor.

Whereas an inductor opposes any change in current flow, a capacitor, on the other hand, blocks any change in voltage. As such, the current through a capacitor is a function of voltage change, which is given as follows:

$$i_c = C\frac{dv}{dt} = C\left(2\pi f V_p \sin\left[2\pi ft + 90^o\right]\right). \tag{1.9}$$

Because of the electrostatic field around a capacitor, opposition to the flow of AC is created. This is called capacitive reactance (X_c), computed as

$$X_C = \frac{1}{2\pi fC}, \tag{1.10}$$

where, f is the frequency of the AC source (Hz) and
C is the capacitance in farad (F).

The combined opposition to the flow of AC current in a circuit is called impedance (Z). This is a summation of the resistance, capacitive reactance, and inductive reactance present in a given circuit. It is computed as

$$\begin{aligned} Z &= R - j\frac{1}{2\pi fC} + j2\pi fL \\ &= \sqrt{\left\{R + (X_L - X_C)^2\right\}} \end{aligned} \tag{1.11}$$

The j-notation is used because the inductive and capacitive reactances are plotted in the imaginary axis of a complex plane, as will be seen later.

Resistors and inductors in series are summed up by adding individual elements:

$$\begin{aligned} R_{eq} &= R_1 + R_2 + \cdots + R_N, \\ L_{eq} &= L_1 + L_2 + \cdots + L_N. \end{aligned} \tag{1.12}$$

For capacitors connected in series, the summation is done using the inverse summation as follows:

$$\frac{1}{C_{eq}} \equiv \frac{1}{C_1} + \frac{1}{C_2} + \cdots + \frac{1}{C_N}. \tag{1.13}$$

For two or more elements that share a common node, the sum of resistors and inductors is given as follows:

$$\frac{1}{R_{eq}} = \frac{1}{R_1} + \frac{1}{R_2} + \cdots + \frac{1}{R_N},$$

$$\frac{1}{L_{eq}} = \frac{1}{L_1} + \frac{1}{L_2} + \cdots + \frac{1}{L_N}. \tag{1.14}$$

For two or more capacitors connected in parallel, the equivalent total is the algebraic summation of the individual units:

$$C_{eq} = C_1 + C_2 + \cdots + C_N. \tag{1.15}$$

Capacitive and inductive behavior in AC circuits affect the computation of power delivered and utilized in the said circuits. The product of voltage and current (VA) is called apparent power. Apparent power is the power drawn by a given electrical equipment. The actual power utilized is called true power and is measured in watt. Reactive power is the component of power that flows across the capacitor or inductor and is measured in volt-ampere reactive (vars). Reactive power is used to supply energy that oscillates in the magnetic or electric field. The ratio of true power to apparent power represents the conversion efficiency of a given load and is called a power factor. This is shown in the power and impedance triangle in Figure 1.14.

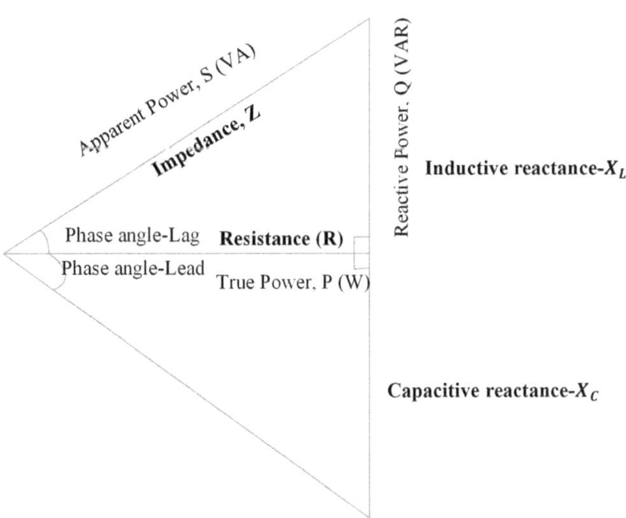

FIGURE 1.14 Combined Power–Impedance Triangle.

Current in an inductive load lags the applied voltage by a certain angle, as seen in Figure 1.7. As such, an inductive load will have a lag power factor. Examples are transformers, motors, and many residential and industrial loads.

However, for a leading power factor, the current leads the applied voltage as exemplified by capacitive loads.

In a nutshell:

- Real power does the actual work of heat dissipation, lighting, and mechanical works such as running of motors.
- Reactive power supports magnetic and electric fields that run AC systems.
- The summation of the above two is called apparent power, which is measured in volt-amps (VA).
- The efficiency of conversion of the power delivered to useful power is called the power factor. It is the ratio of real power divided by apparent power.

1.3 CHAPTER SUMMARY

This chapter started with a look at the methods of generation of electric energy. After that, the major technological discoveries that had far-reaching implication on the evolution of power systems were discussed. This included aspects such as the advent of AC and HVDC power systems and the modernization of the same over the years.

The chapter further looked at the development of power system standards, grid codes, and sector liberalization many years later vis-à-vis rapid technological advancement in the age of smart grid and the 4IR. This was followed by a peep into the possible future of wireless power transmission and the impact of Society 5.0 on power systems.

Finally, the chapter looked at the basic power system concepts in the areas of circuit network, theorems, and basic network analysis.

REFERENCES

[1] MADE EASY Publications, *Power Systems: Comprehensive Theory with Solved Examples and Practice Questions*, 4th ed., MADE EASY Publications, 2018.

[2] F.G. Longatt, *Introduction to Power Systems*, Researchgate, December 2019.

[3] M. Brimhall and D. Crane, *Grid Fundamentals*, WECC, 2021.

[4] O.P. Malik, "Evolution of power systems into smarter networks," in *Journal of Control, Automation and Power Systems*, vol. 24(112), pp. 139–147, April 2013.

[5] CRC Press, *Electrical Power Systems Technology*, 3rd ed., CRC Press, 2008.

[6] A.V. Meier, *Electric Power Systems: A Conceptual Introduction*, IEEE Press, 2006.

[7] D.P. Kothari and I.J. Nagrath, *Modern Power System Analysis*, 3rd ed., Tata McGraw Hill Education Private Limited, 2006.

[8] L. Philipson and L. Willis, *Understanding Electric Utilities and Deregulation*, 2nd ed., CRC Press, Taylor & Francis, 2006.

[9] S. Fardo and D. Patrick, *Electrical Power Systems Technology*, 3rd ed., CRC Press, 2008.

[10] M. El-Hawary, *Electrical Energy Systems*, CRC Press, 2000.

[11] P. Ponce et al., *Power System Fundamentals*, CRC Press, 2018.
[12] P. Kundur, *Power System Stability and Control*, McGraw-Hill Inc., 2000.
[13] D. Glover et al., *Power System Analysis and Design*, Cengage Learning, 2012.
[14] M. Ceraolo and D. Poli, *Fundamentals of Electric Power Engineering*, IEEE Press, 2014.
[15] J. Casazza and F. Delea, *Understanding Electric Power Systems: An Overview of the Technology and the Marketplace*, IEEE Press, 2003.
[16] S. Karris, *Circuit Analysis I with MATLAB Applications*, Orchard Publications, 2004.
[17] J. Fiore, *AC Electrical Circuit Analysis: A Practical Approach*, Dissidents, 2023.
[18] M. Mbae and N. Nwulu, "Filter feeding allogenic engineering optimization algorithm for economic dispatch," in *International Journal of Energy Production and Management*, vol. 6(2), pp. 113–128, 2021.

2 Evolution of Power Transmission and Distribution Systems

2.1 POWER TRANSMISSION LINES

A transmission line is the interconnector between power generation and power distribution systems. Any power transmission system has important parameters, namely, resistance, inductance, capacitance, and conductance.

The resistance and inductive reactance (summed up to form series impedance) are uniformly distributed along the length of the transmission line. Capacitance is created between two conductors separated by an insulating material, e.g., air, for transmission lines.

For the purpose of line modeling and carrying out design and simulations, transmission lines are broadly classified as

- short-length transmission lines (less than 80 KM long, less than 66 KV)
- medium-length transmission lines (80–250 KM long, 66 KV–100 KV)
- long transmission lines (more than 250 KM long, more than 100 KV) [1–5].

2.1.1 Short-Length Transmission Lines

Because of the short length of the line and the applied low transmission voltage, the capacitance is small, and hence neglected for the purpose of line modeling. As such, the equivalent circuit is as seen in Figure 2.1.

I_s and I_R are the sending and receiving end currents, respectively, and V_s and V_R are the sending and receiving end voltages, respectively. The line is approximated as a simple series circuit using Equation 2.1:

$$V_S = V_R + I_R Z. \tag{2.1}$$

2.1.2 Medium-Length Transmission Lines

The line is sufficiently long besides the relatively high amount of transmission voltage; thus, there is enough capacitance that needs to be factored when developing the line model.

To make computations simple, the line capacitance is usually lumped at one or more points along the line, either as end condensed, nominal T, or nominal. Figure 2.2 represents nominal π.

DOI: 10.1201/9781032665290-2

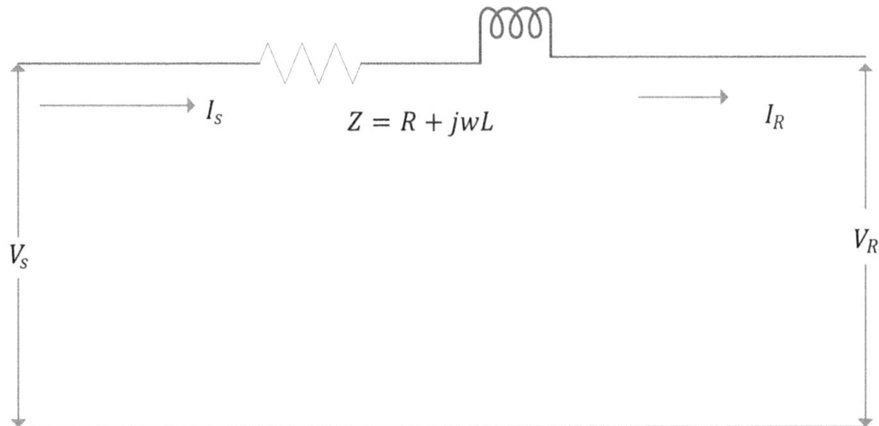

FIGURE 2.1 Equivalent Circuit of a Short Transmission Line.

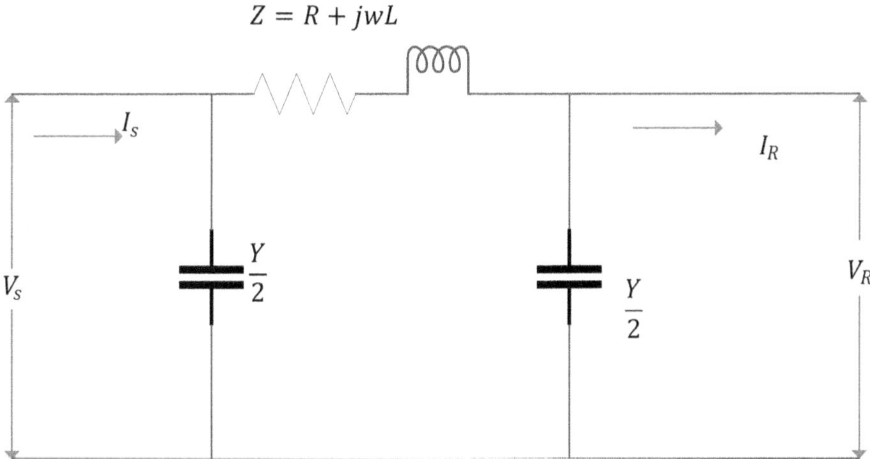

FIGURE 2.2 Equivalent Circuit of a Medium Transmission Line.

The total shunt admittance is divided into two and placed at the respective ends of the line as seen above.

The sending end voltage is given by Equation 2.2:

$$V_S = \left(V_R \frac{Y}{2} + I_R\right) Z + V_R = \left(\frac{ZY}{2} + 1\right) + ZI_R. \tag{2.2}$$

The sending end current is given by Equation 2.3:

$$I_S = V_S \frac{Y}{2} + V_R \frac{Y}{2} + I_R = {}_R Y\left(1 + \frac{ZY}{4}\right) + \left(\frac{ZY}{2} + 1\right) I_R. \tag{2.3}$$

2.1.3 Long Transmission Lines

The analysis is done using the fact that the line parameters are uniformly distributed on the line. As such, the impedance and admittance are uniformly distributed along the length of the line. Granular differential equations are used to model the line parameters, as shown in Figure 2.3.

Given z = series impedance per unit length,
y = shunt admittance per unit length,
l = length of the line,
$Z = zl$ = total series impedance,
$Y = yl$ = total shunt admittance.

For analysis purposes, we will take an elemental length Δx at a distance x from the receiving end of the line. If the voltage and current at distance x are V and I, respectively, the same at $x+\Delta x$ will be $V+\Delta V$ and $I+\Delta I$, respectively. As such,

$$\Delta V = Iz\ \Delta x \rightarrow \frac{\Delta V}{\Delta X} = Iz,$$

$$\Delta I = Vy\ \Delta x \rightarrow \frac{\Delta I}{\Delta x} = Vy. \tag{2.4}$$

For the limit $x0$, Equation 2.4 becomes

$$\frac{dV}{dx} = Iz,$$

$$\frac{dI}{dx} = Vy. \tag{2.5}$$

Differentiating 2.5 gives

$$\frac{d^2V}{dx^2} = z\frac{dI}{dx} = zyV \rightarrow \frac{d^2V}{dx^2} - zyV = 0. \tag{2.6}$$

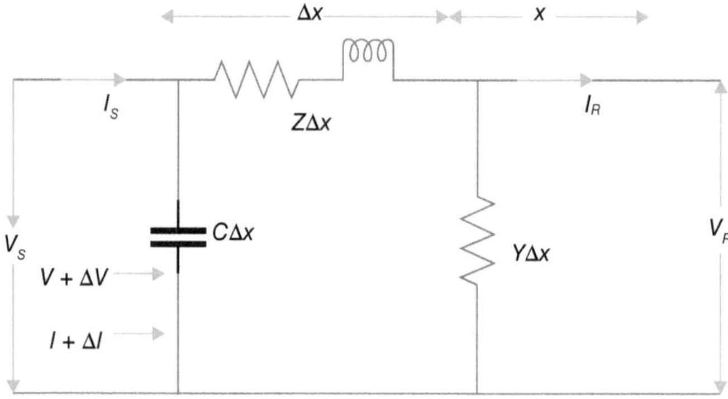

FIGURE 2.3 Model of a Long Transmission Line.

This differential equation is solved to give

$$V = Ae^{x\sqrt{zy}} + Be^{-x\sqrt{zy}}. \tag{2.7}$$

The work is to determine constants A and B for given boundary conditions and given that the receiving end voltage and currents are known.

2.2 VIRTUAL POWER PLANTS

A virtual power plant (VPP) refers to an aggregation or combination of diverse and decentralized resources such as generation, storage, and controllable variable loads in a coordinated manner to deliver grid services and customer and consumer benefits. A good example is the aggregation of rooftop solar generation with battery storage and controllable variable loads, such as space heating and cooling, and pumping systems, among others. This involves the use of advanced software and communication tools to deliver services that are ideally done by a convection power generation plant. Cloud-to-cloud communication and control technology are gaining prominence in this area [6, 7].

The illustration in Figure 2.4 shows a VPP model used to aggregate battery storage produced by a number of residential houses within a given residential neighborhood.

A power system operator treats a given VPP as a single power generation plant generating a given amount of power for all operational purposes. This is the essence of the use of the term virtual plant, as basically the VPP is seen as a single plant, albeit a virtual one as opposed to the conventional power generation plant.

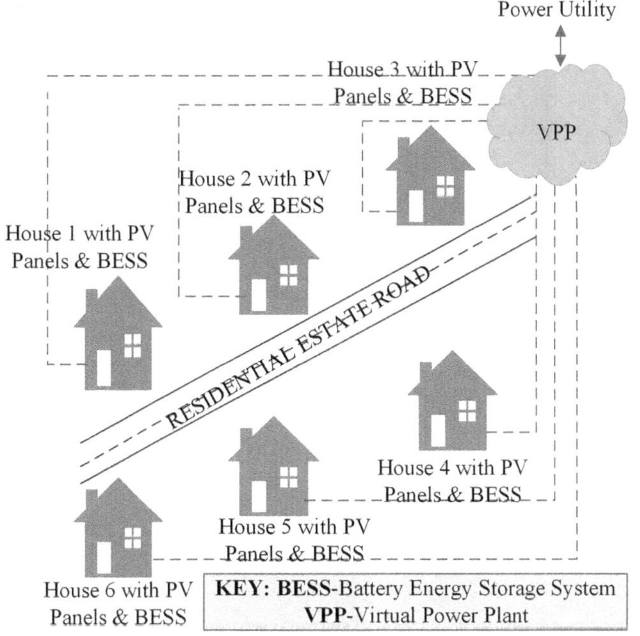

FIGURE 2.4 DER VPP Aggregation System Architecture.

They have the capability to provide customized services in terms of reliable and affordable energy for different circumstances and requirements of diverse prosumers. The overall objective will be to optimize aggregated large-scale clean distributed energy resources (DER) and complex power systems that deliver value for the entire value chain, both in the wholesale and retail markets. Power systems are capital-intensive investments; thus, there is a great need to optimize the building and operation of the power infrastructure. Deferring large-scale generation asset investment goes a long way in improving a power utility's cash flow. Consumers seek to get value for their investment in generation, as utilities seek the available generation for frequency control among other grid and contingency services. VPPs have been shown to increase frequency generation in the event of loss of a major generator(s) or transmission line(s) in a utility.

An operational VPP architecture is as illustrated in Figure 2.5.

A large number of batteries operated in unison form a VPP that acts as a large power plant that delivers services through the existing transmission and distribution network, subject to operational constraints. The services include but not limited to dispatching energy in case of major generator or transmission line outages, peaking services, grid ramping service, transmission network congestion easing, and given frequency control ancillary services.

Synchronous integration of VPPs into the system in a planned manner is critical to avoid instantaneous overload of existing transmission and distribution networks [8–11].

As an example, Australia has a fairly advanced VPP program. Good examples include the following:

- Next Generation Energy Storage Program, installing battery storage in more than 5,000 Canberra homes and creating a 36-MW VPP capability.

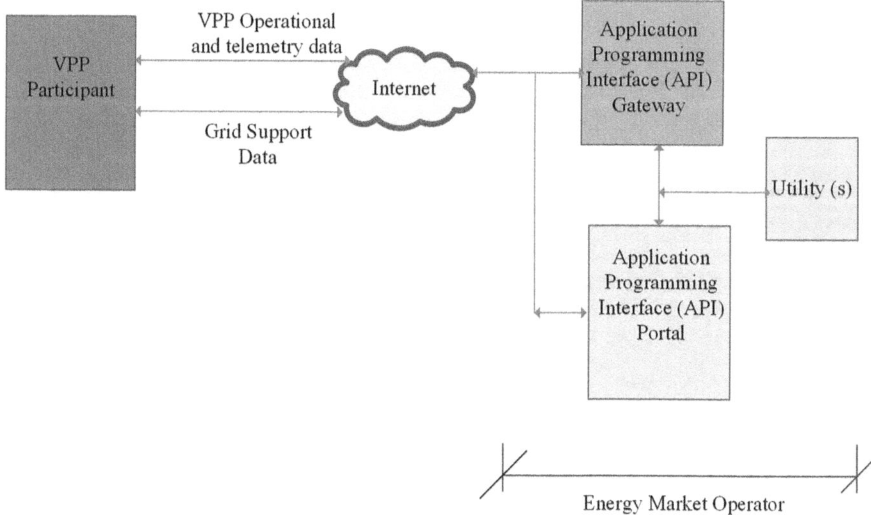

FIGURE 2.5 DER VPP Aggregation System Architecture.

- Simply Energy's 8-MW VPP to be developed in Adelaide, comprising 6 MW of residential battery storage and 2 MW of demand response capacity from commercial businesses.
- The South Australian government's USD 100 million Home Battery Scheme to incentivize 40,000 new battery storage systems [7].

The legal and regulatory framework will need to evolve to take care of the emerging operational environment. VPPs are only able to work as aggregators of distributed generation resources once the consumers allow them control over their generation equipment and load. Consumer rights, contractual obligations, and expectations, and charging and discharging times vis-à-vis their energy usage patterns need to be brought out at the onset to all prosumers. Necessary incentives such as subsidy on the cost of panels and storage, and a competitive, granular tariff, among others, will go a long way in encouraging many customers to sign in. The challenge and risk in managing peak loads will have to be factored in to better encourage customers to board.

As the VPPs move toward a phase of rapid growth, a number of things that need to be addressed include but are not limited to the following:

- Flexibility in the charging and discharging time of batteries.
- Versatile contracts for panel cleaning and battery maintenance.
- A single program for a county/state/country/region to achieve economies of scale.
- Competitive tax regime, rebates, incentives, and financial and technical assistance for joining prosumers.
- More incentives for environmental protection gains.
- Use of smarter and more efficient appliances.
- Legislate for all upcoming developers to install solar panels and storage in their buildings.

There are still no widely accepted international standards for DER to be connected to a power system as a VPP. At a minimum, the storage system has to have the following:

- Battery modules
- An inverter system
- Telemetry capacity
- Two-way communication system
- Smart controller
- Smart meter
- PV panels.

The communication system should allow real-time system monitoring on aspects such as battery state of charge, battery real and reactive power, system coupling point voltage, and real-time system control, among other requirements for a robust virtual power plant.

The operational architecture of a VPP is illustrated in Figure 2.6.

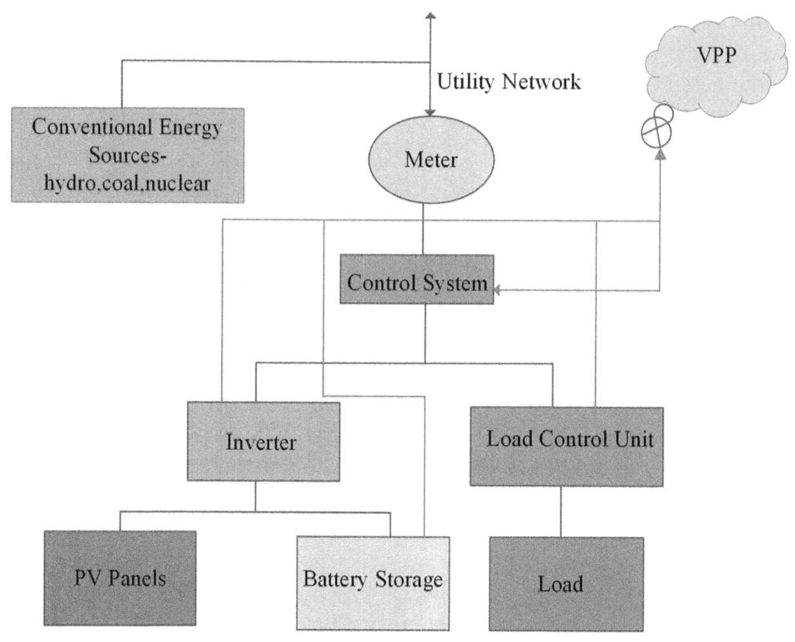

FIGURE 2.6 Layout of a VPP System.

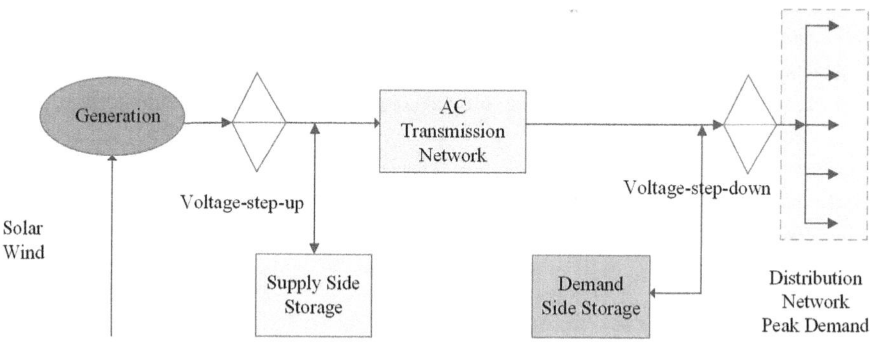

FIGURE 2.7 Layout of a Virtual Transmission Line.

2.3 VIRTUAL TRANSMISSION LINES

Virtual transmission lines are made up of battery storage connected to the power grid on the generation side (to store excess renewable generation that cannot be transmitted due to grid congestion) and on the distribution side (to be charged when grid congestion subsides and discharged on a need basis) [6, 7].

This is illustrated in Figure 2.7.

The supply-side scalable storage charges to store the generated renewable energy, and the demand-side storage charges when network capacity is available. Demand-side storage discharges to address peak demand needs when there is grid congestion for the network between generation and load sides. It is a cheap way of supporting the grid to address certain seasonal peak loads, e.g., at the peak of cold or hot seasons. The lines also assist power system operators to address the challenge of redispatch, where in many cases, the utility has to pay for the generation, whether taken up or not.

Virtual transmission lines allow deeper integration of variable renewables, such as wind and solar power, with reduced network congestion, thus allowing for postponing of expensive network expansion projects. The said upgrades include more overhead lines, increasing the current-carrying capacity of existing lines and cables, and adding new or expanding existing primary substations.

Battery storage can also support the grid by offering ancillary services such as frequency control, voltage regulation, and spinning reserve.

Battery storage can be scaled in terms of size and plugged in or out of the grid on a need basis. As such, they offer a big advantage in terms of usage flexibility, planning, operation, and deployment over different time horizons. Unlike the convection grid, batteries can be located at a variety of points on the grid, with a potential of being moved to a new location as need arises.

2.4 WIRELESS POWER TRANSMISSION

The convection power system uses conductors for power transmission and distribution. This type of power system has the drawback of aging, wear and tear, impact of vagaries of adverse weather, and terrain challenges, among others. Wireless power transfer is a promising concept toward addressing a number of these challenges. This is illustrated in Figure 2.8 [12–16].

This was first demonstrated by Nikola Tesla in 1893 when he lit a vacuum bulb using energy that was delivered wirelessly.

One option is to use microwave line-of-sight transmission. The process starts with the conversion of electric power to microwaves, which are then broadcast by the transmitter. The final receiver converts the waves back to DC and then to the AC waveform. The principles of electromagnetic induction come into play here.

The transmitter can also convert the electric current to a laser beam, which is sent to the receiver (rectenna). The receiver has photovoltaic cells that convert the laser light back to AC current. Again, this also requires line-of-sight linking between the transmitter and receiver. This is in addition to the huge energy losses in the entire process.

Satellites coupled with huge solar arrays have also been used as a medium to transmit electric power between transmitters and receivers.

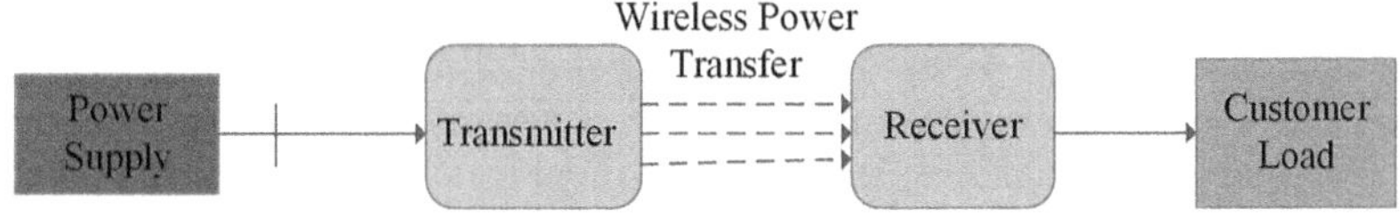

FIGURE 2.8 Wireless Power Transfer.

In all the cases, interference by atmospheric conditions such as rain, fog, clouds, and storm presents a huge challenge to efficient power transmission.

Currently, we need to use very large transmitters to achieve any reasonable amount of energy transmission. This implies a high cost of the system. As the technology matures and becomes more efficient, (microwave) wireless power transfer promises to be more cost-efficient, versatile, and flexible means of power transmission and distribution. The system is free from the line faults experienced in the conventional power systems, e.g., line-to-line or line-to-ground faults. Power theft by tapping the along the line should also be eliminated subject to proper system protection.

The challenge of efficiency has been partly addressed by the use of single-wave patterns to form a collimated beam, the waves of which are totally aligned in parallel. A series of relays used between the transmitter and the rectenna goes a long way toward creating a sharp focus for the transmitted beam, as well as reducing transmission losses. Most of the losses occur at the rectenna conversion. A commonly used rectenna is the series-parallel assembly of Schottky diodes.

The performance of wireless power transfer systems can be increased by the use of several antennas configured so that successive units combine their fields at the rectenna (array or antenna gain). Due to rectenna nonlinearity, a beam steering algorithm helps optimize power transfer in real time.

Several standards to govern wireless power transfer, namely, IEC 61980 and ISO 19363, among others, are at various stages of development.

To achieve the required safety levels, a nonionizing industrial, scientific, and medical frequency band is used.

As wireless power transfer takes the center stage, the level of utilization will grow in leaps and bounds. The range of the electromagnetic waves will be much larger. At an average range of 20 m, entire homes will be able to be charged by a single transmitter located close in the vicinity. As the range grows, entire blocks, streets, and neighborhood will be powered by a single rectenna. Once this point emerges, consumers will have full electric mobility, with all cars wirelessly charging. Roads will adapt to having wireless chargers spaced far apart so that a car can run endlessly. Application of wireless power transfer for the rail network, aerial unmanned vehicles, wireless sensors, and Internet of Things (IoT) devices for research, data collection and other purposes will greatly revolutionize the process of power generation, transmission and end use.

Currently, radio frequency has a fairly low power density when compared to other ambient energy harvesting technologies. The future will be in frequency bands that will enable simultaneous wireless power and data transmission. The ambient energy sources are still not very reliable due to low equipment efficiency, interference of climatic conditions, and their variable nature. However, once mature, they hold the promise of post-grid, battery-less clean power systems.

The ultimate frontier in this direction will involve solar panels in outer space to collect the Sun's energy, whereby solar-powered satellites will beam it back to the Earth via microwave power transmission. This promises a limitless supply of energy. This will take years to achieve, but the world should start to look at this as the solution of choice. Wireless power transfer will solve the world's energy crisis.

Metamaterials have been able to achieve a high degree of transmitter efficiency—upward of 75% (good but less than what conductors like copper and aluminum are able to achieve). Metamaterials are nanotechnology engineered heterogeneous materials designed to have special manipulation of electromagnetic and light waves—nanoscale optical antennas. They are designed to block, enhance, absorb, or bend the electromagnetic waves.

They are grouped into resonant and nonresonant metamaterials. Resonant metamaterials have meta-atoms in their resonant state, resulting in high values of negative electrical permittivity and magnetic permeability, which, in turn, enables achievement of negative values of refractive index. They have high losses, very short wavelengths, and a narrow bandwidth. Nonresonant metamaterials come on board to address the issue of large losses. This is because their meta-atoms are far from the resonant state, resulting in steeped electromagnetic response.

Metamaterials are anisotropic, giving them the ability to control medium parameters, thus improving their ability to control electromagnetic waves.

Metamaterials are employed in many optical transformations and photonic applications such as antenna and radar construction, precise telecommunication antennas, lasers, defense camouflage surveillance, medical imaging, high-resolution optical lenses and filters, subwavelength imaging, and energy harnessing at low energy intensity such as in the rectenna and in smart harnessing and management of solar energy.

The continuous development and research into metamaterials are a multidisciplinary activity that encompasses diverse fields such as classical optics, material science, nanotechnology, additive manufacturing, quantum mechanics, solid-state physics, electromagnetics, microwave and antenna physics, and semiconductor physics, among others.

Research and development of better quality metamaterials will go a long way to making wireless power transmission an efficient venture. The future looks bright in this regard.

2.5 POWER TRANSMISSION AND DISTRIBUTION SYSTEMS OF THE SMART GRID

A smart or intelligent grid is a versatile power system that allows the two-way flow of energy and is managed by an advanced communication and control system. This is illustrated in Figure 2.9 [8].

The grid allows a range of functions, applications, and capabilities that are ordinarily not possible in a conventional grid. The switch to smart grid is fueled by a number of compelling needs that include, but are not limited, to the following:

- Improvement of grid reliability in terms of real-time balancing of supply and demand by handling generation, transmission, and load disruptive events such as during major system faults.
- Better grid resilience to ensure continued operation in the face of adverse faults and breakdowns caused by storms, floods, etc. This refers to the ability

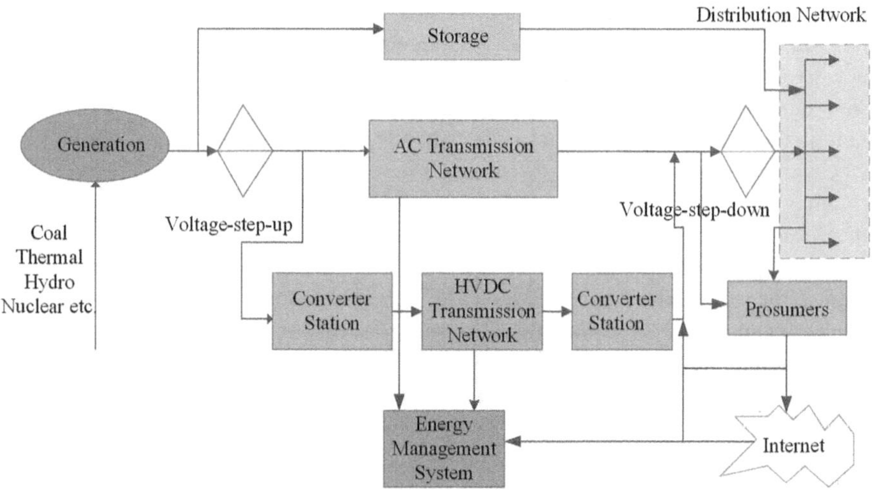

FIGURE 2.9 Smart Grid.

of the grid to bounce back at the shortest time possible after the occurrence of catastrophic events.

- Improved grid security through real-time system monitoring and control to ensure that the system operates as close as possible to the defined normal operating state. This includes activities such as reactive power generation and absorption to ensure that voltage levels are within the defined limits.
- Better grid operational flexibility as a result of increased system integration of variable renewables such as wind and solar power enables real-time balancing of demand and supply. This ensures better grid stability and reliability.
- Deeper penetration of renewable energy, distributed generation, and storage. Real-time two-way flow and control of energy is a critical ingredient toward management of the rising tide of prosumers.
- Enabling the rise of VPPs that enable the aggregation of distributed renewable energy generation within a given locality. The smart power system is able to treat the said generators as one big power plant.
- Reduction of technical and commercial losses for better utility sustainability. Continuous and real-time data analytics enable pinpointing of areas of commercial and technical constraints that require action.
- Improved asset integration management by means of optimization on low cost generation, optimal usage of transmission and distribution assets, and data analytics for predictive and reliability-centered asset maintenance.
- Adoption of granular tariff such as the electric mobility and time-of-use tariff to enable, e.g., industrial producers to take advantage of cheap power during off-peak time.
- Reduction of the electric carbon footprint by deepening the integration of clean renewable energy technologies into the grid.

A smart grid needs the integration of smart sensors, automated and smart switches and circuit breakers, smart meters, and communication and data processing software among other key components to enable real-time system operation and control. The grid is actually the convergence between physical assets of the power system and communication technologies. This is one of the central themes of the Fourth Industrial Revolution (4IR).

A multiplicity of smart sensors, robust communication systems, and telemetry allows real-time prediction, detection, and action to sort out system faults. This allows the system to be very resilient in the face of major generation and transmission system breakdowns; natural disasters such as floods, hurricanes, and storms; accidents; and other forms of system disturbances. This is achieved by self-healing, by which faulted parts are isolated to prevent cascaded outages and further spread of damage. Real-time network reconfiguration assists to minimize the number of affected customers.

The smart grid is a critical requirement for the onboarding of consumers for participation in the wholesale and retail electricity markets. Smart grid helps to continuously gather and share useful information to prosumers for better decision making as active participants in both energy generation and consumption. The IoT is a good example where customers have full control of their energy usage and can grant utilities control over their assets as exemplified in VPPs.

Table 2.1 shows the key differences between smart and conventional power grids.

Maintaining power quality by ensuring that voltages are within legal limits is one of the key uses of a smart grid. All the key system buses, lines, and customer connection points are continuously monitored for possible injection or absorption of reactive energy (VAR) from reactors, capacitor banks, and flexible AC transmission system (FACTS) devices among other ways of voltage–reactive power control. This ensures a high power factor that minimizes technical power losses.

This continuous monitoring and control of the smart power system is one of the functions of the supervisory control and data acquisition (SCADA) system. SCADA is part of a smart grid. It enables real-time monitoring and control of various power system buses, lines, and points. It is composed of a number of components including,

TABLE 2.1
Differences between Conventional and Smart Grids

Smart Grid	Conventional Grid
Has digital equipment	Uses electromechanical devices
Two-way communication	One-way communication
Two-way flow of power	One-way flow of electric power
Uses smart sensors	Few, if any, sensors
Smart metering	Electronic/electromechanical metering
Real-time system monitoring	Blind system monitoring
Self-healing capability	Manual system restoration
Remote system operation	Manual system operation

but not limited to, smart sensors and telemetry equipment, smart switches and circuit breakers, a robust communication system, a good software platform, and computer capability. A SCADA system can perform the following actions:

- Control and operation of switches and circuit breakers in the system;
- Operation of reactive power sources and sinks such as capacitor banks;
- Transfer of loads between primary substations and transmission lines;
- Sectionalizing faulted buses;
- Acquisition and processing;
- Automatic line closure after transient faults;
- Reporting the actual location of faults on power lines;
- Reporting on system alarms according to set parameters.

The complete adoption of the smart grid has a number of bottlenecks to overcome. These include the following:

- Development of widely accepted operational and interoperability standards;
- High cost of deployment;
- High cybersecurity risk;
- Lack of sufficient generation and transmission capacity;
- Changing customer needs and expectations;
- Customer privacy concerns;
- Conservative policy, legal, and regulatory space.

2.6 CHAPTER SUMMARY

This chapter looked at the conventional analysis of power transmission and distribution lines by considering the parameters of resistance, capacitance, and inductance. This forms the foundation upon which all advances in the design and construction of transmission lines are anchored. As such, this was a reasonable introduction as the chapter progressed to VPPs and virtual transmission and distribution lines. This gave way to the discussion of wireless power transmission networks, which is a rapidly growing area.

Lastly, the progression to the smart grid in comparison to the conventional grid was discussed.

REFERENCES

[1] T. Gonen, *Engineering of Electric Power Transmission Systems*, 3rd ed., CRC Press, 2014.
[2] U.A. Bakshi and M.V. Bakshi, *Power System-I*, 1st ed., Technical Publications, 2009.
[3] J. Grainger and W. Stevenson, *Power System Analysis*, 1st ed., McGraw-Hill Inc., 1994.
[4] D. Glover and T. Overbye, *Power System: Analysis and Design*, 5th ed., CENGAGE Learning, 2012.
[5] P. Ponce et al., *Power System Fundamentals*, CRC Press, 2018.
[6] Australia Energy Market Operator (AEMO), *AEMO Virtual Power Plant Demonstration*, AEMO, 2020.

[7] P. Van, *Australia Energy Market Operator (AEMO)*, Virtual Power Plant Demonstrations Consumer Insights Report, 2021.
[8] T. Flick and J. Morehouse, *Securing the Smart Grid; Next Generation Power Grid Security*, Elsevier, 2011.
[9] C.L. Wadhwa, *Electrical Power Systems*, New Academic Science, 2012.
[10] D. Das, *Electrical Power Systems*, New Age International Publishers, 2006.
[11] T.K. Nagsarkar and M.S. Sukhija, *Power System Analysis*, 2nd ed., Oxford University Press, 2014.
[12] S. Kumar, "Introduction to wireless power transmission," in *International Journal of Technology Innovations and Research*, vol. 8, pp. 1–10, 2014.
[13] A.M. Jawad et al., "Opportunities and challenges for near-field wireless power transfer," in *MDPI Journal*, vol. 10, 2017.
[14] B. Zhu and X. Gao, "Review of magnetic coupling resonance wireless energy transmission," in *International Journal of u- and e- Services, Science and Technology*, vol. 8(3), pp. 257–272, 2015.
[15] H. Hong and D. Yang, "The analysis for selecting compensating capacitances of two-coil resonant wireless power transfer system," in *Proceedings of the 1st IEEE International Conference on Energy Internet*, ICEI, pp. 220–225, 2017.
[16] Y. Zhang and Z. Yan, "An LCL-N compensated strongly coupled wireless power transfer system for high power applications," in *IEEE Application in Power Electron Conference Expo*, pp. 3088–3091, 2019.

3 Metamorphosis of Power System Control and Optimization Techniques

3.1 IMPORTANCE OF POWER SYSTEM CONTROL AND OPTIMIZATION

As power networks become larger and more complex, their operation, stability, and control become an evolving challenge that power system operators, practitioners, and researchers continually grapple with.

As global economies expand, power demand is snowballing. As energy generation expands, the end-user quality of supply challenge continues to grow in tandem. The long distances between generators and major industrial, commercial, and residential load hubs only compound the power system stability issue. Several national and regional power outages have been documented over recent years across the world due to increased stress on the existing power network. This presents researchers with opportunities to tackle the emerging challenges in a better, more robust, sustainable, and flexible manner.

The conventional power system stability and control improvement methods such as reactors, capacitors, var systems, and condensers have many inherent drawbacks. The main drawback is that these methods and devices are largely electromechanical systems with challenges related to giving the needed high-speed control and flexibility.

FACTS devices have presented a versatile solution toward the handling the varying operational conditions and complexities to improve power system stability, transfer capability, and control. Their main advantages over the conventional methods are that the FACTS devices are incredibly dynamic, versatile, and flexible [1–10].

3.2 POWER SYSTEM OPTIMIZATION

3.2.1 Power System Stability and Control

The stability of a power system refers to the ability of the power system to maintain a balanced state under normal conditions and upon the occurrence of small to severe disturbances in the system [3].

The stability of the power system is classified mainly into two, as illustrated in Figure 3.1.

DOI: 10.1201/9781032665290-3

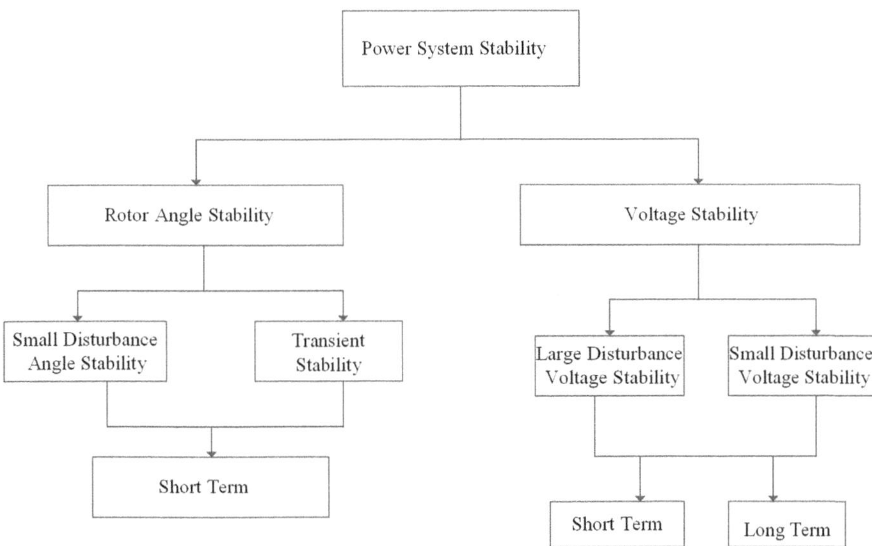

FIGURE 3.1 Classes of Power System Stability.

Voltage stability is the attribute of a power system that possesses the required voltage magnitude at all buses under normal system conditions and upon the occurrence of a disturbance. The challenge of supplying enough reactive power at the point where it is needed is the leading cause of voltage instability. This challenge is caused by very big loads, having voltage sources very far from the load hubs, low voltage sources, and low reactive power compensation levels of the load, among other causes.

Several sources and sinks of reactive power are used in a power system for voltage control and regulation. These include, but are not limited to, synchronous generators, loads, overhead lines, underground cables, transformers, shunt and series capacitors, reactors, static var systems, and synchronous condensers.

The stability of the rotation angle is the characteristic of a power system that makes it remain in step after being subjected to electromechanical oscillations. It can be improved by using several devices, including high-speed breakers for fault clearance, series capacitor banks, synchronous compensators, static var compensators, resistors, and shunt reactors.

Most of the abovementioned methods of a power system are largely mechanical devices that lack the high-speed control and versatility required in a real power system. FACTS devices provide a tool to offer the much-needed high-speed control and flexibility for modern and increasingly complex power systems. They provide an excellent tool for improving power system control, stability, and power transfer capability.

The huge capital investment required for the new power infrastructure, land scarcity, and pressure from environmentalists have necessitated the use of alternative methods of power systems to optimize the existing infrastructure. Again, FACTS devices come in handy in this regard. However, their optimal placement and operation for given system constraints are critical due to their high initial installation and maintenance costs.

FACTS devices are power electronic devices that control power system parameters to improve system control and better power transfer capacity. This happens through appropriate adjustment of the three variables of voltage, angle, and impedance. FACTS devices can be used under steady, transient, and post-transient conditions.

They are classified using criteria that include the following:

- Mode of connection to the power system
- Age of FACTS devices
- Switching technology [11–20].

These are illustrated in Figures 3.2–3.5.

FACTS devices are installed to improve power delivery and control the bus voltage in a power system for better stability. Series FACTS devices adjust the line impedance to increase the amount of power transmitted. On the other hand, shunt FACTS devices work as a controllable current source.

They have been instrumental in making the best use of the existing power network and, in essence, deferring the need for new infrastructure. This is in addition to growing land constraints and growing environmental concerns.

Due to their high initial installation and maintenance costs, placement of the said devices in the best location in a power system is critical. Continuous improvement of placement techniques for existing FACTS devices and the development of new placement methods continue to attract a lot of research interest.

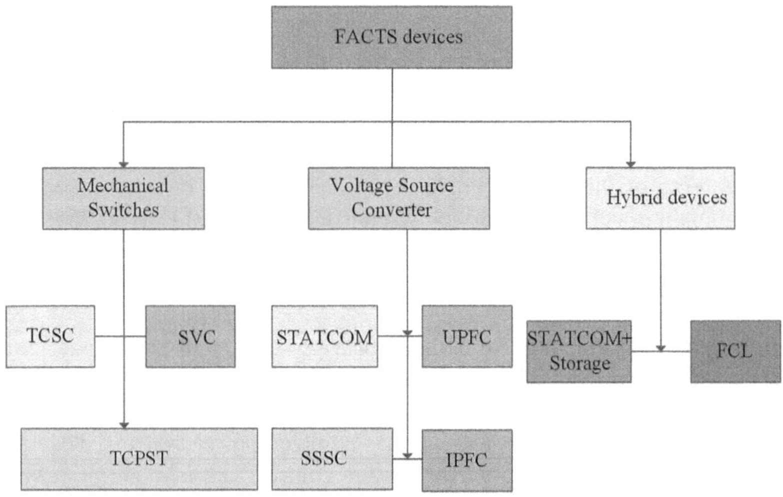

FIGURE 3.2 Classification of FACTS Devices Using Switching Technology.

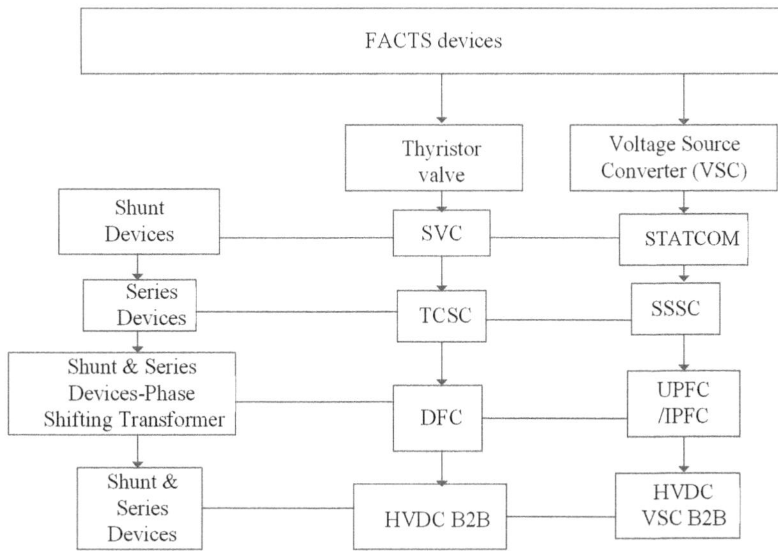

FIGURE 3.3 Classification of FACTS Devices Using Switching Technology and Mode of Connection.

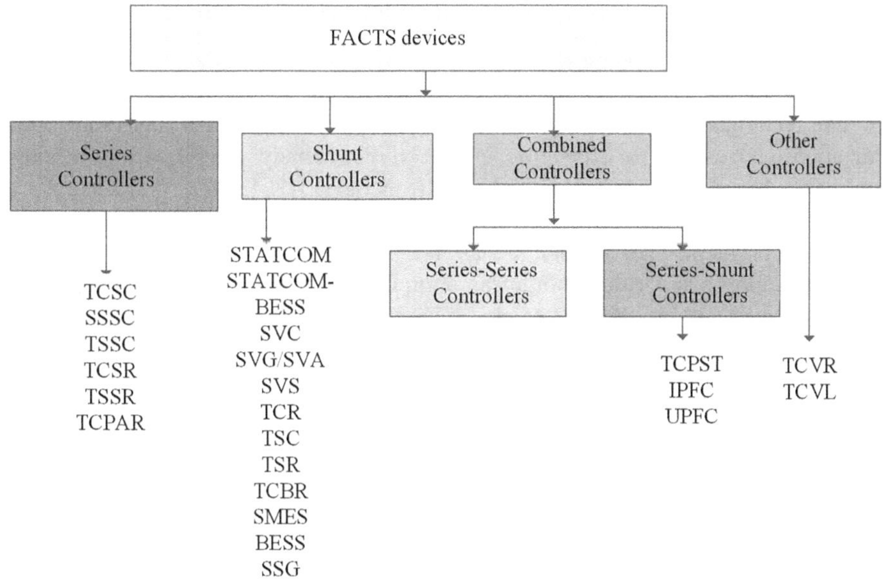

FIGURE 3.4 Classes of FACTS Devices Based on Mode of Connection.

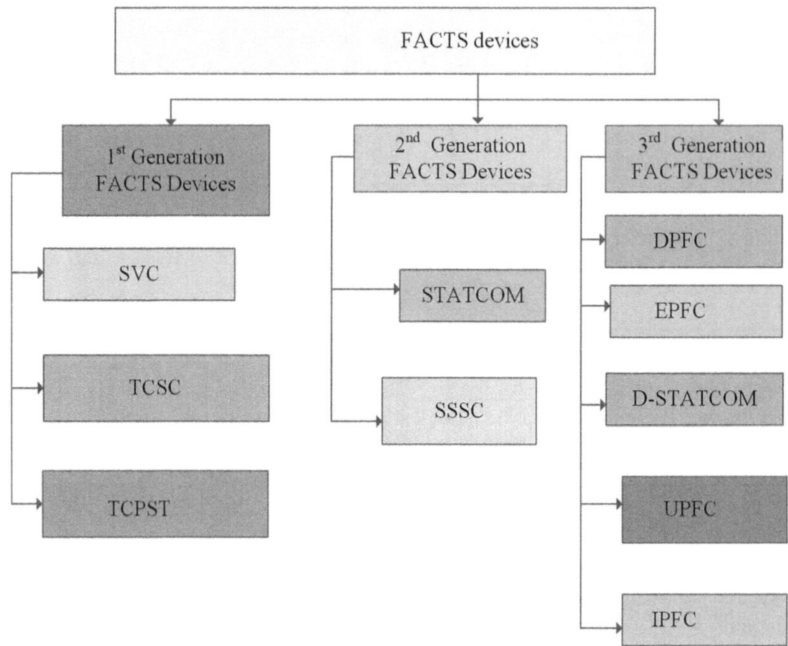

FIGURE 3.5 Classes of FACTS Devices Based on Their Age.

3.2.2 Categorization of Power System Optimization Techniques

Optimization of existing power systems is a growing concern across the world, as the need for deferral of expensive system upgrades, among other challenges, continues to grow.

The techniques for optimal placement of FACTS devices (and any other optimization devices) are broadly grouped into three, namely, the classical methods, sensitivity-based methods, and meta-heuristic methods [12, 13, 18].

Classical optimization methods make use of mathematical equations to solve the placement problem. They include, but are not limited to, curved space optimization, Newton–Raphson algorithm, nonlinear optimization programming, dynamic optimization programming techniques, and mixed integer and mixed-integer nonlinear methods. These methods have several weaknesses, namely, the convergence challenge occasioned by the nonlinear nature of power system equation formulation, the inability to handle many variable constraints, and high computational time requirement.

Sensitivity-based methods are formulated to track the changes in power system output parameters on the continuous variation in the input variables until a set threshold is achieved. They include, but are not limited to, the Jacobian matrix index, the line stability index, the augmented matrix, the reactive power loss sensitivity indices of the system, the voltage stability indices, the eigenvalue analysis, and the real power flow performance index. Their computational efficiency is highly compromised when handling nonlinear, multivariable, and constrained practical power systems.

Meta-heuristic techniques are inspired by human or biological intelligence and laws of nature and physics. They are broadly classified into four categories, as shown in Figure 3.6.

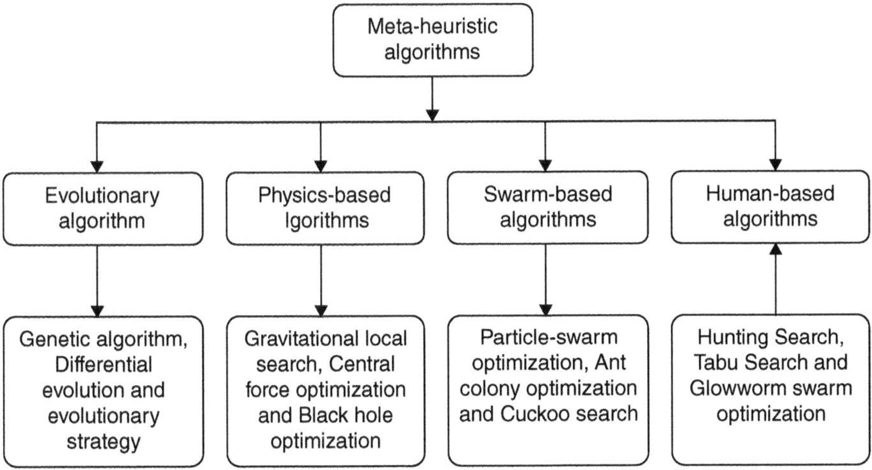

FIGURE 3.6 Meta-heuristic Optimization Methods.

They provide a highly versatile tool for solving multimodal, discrete, stochastic-based, and multi-objective nonlinear constrained power system problems with superior computational efficiency and excellent handling of power system operational constraints. Examples include genetic algorithm, particle swarm optimization, brainstorm optimization, moth–flame algorithm, adaptive cuckoo search algorithm, strength pareto multi-objective evolutionary algorithm, differential evolution, biogeography-based optimization, gravitational search algorithm, fuzzy logic, artificial neural network, harmony search, Grey wolf optimization, hybrid fruit-fly firefly optimization, bee colony algorithm, oppositional krill herd algorithm, flower pollination algorithm, and firefly algorithm [21–25].

3.3 RECENT ADVANCES IN POWER SYSTEM OPTIMIZATION METHODS

As a result of the good performance of the meta-heuristic methods, there has been a lot of energy geared at combining two or more meta-heuristic methods to take advantage of the superior attributes of each of the individual algorithms. Examples are the bat-particle swarm algorithm, the particle swarm gravitational search algorithm, the gravitational search algorithm, the genetic algorithm with ant colony optimization, and tabu search with simulated annealing [26–45].

Despite their superior computational ability, meta-heuristic optimization techniques suffer from a number of operational challenges. These are mainly occasioned by the nonlinear, complex, and stochastic interaction of various meta-heuristic algorithms. The said drawbacks include, but are not limited to, low convergence rate (cuckoo search algorithm), falling into the regional optimum (differential evolution), getting trapped in the local maxima (genetic algorithm, flower pollination algorithm, and firefly algorithm), partial optimism and premature convergence (particle swarm optimization), undefined convergence time (ant colony optimization), need for parameter tuning (bat algorithm and bee algorithm), and high computation time (invasive weed optimization) [46–61].

As such, the continuous development of new nature-inspired optimization techniques and fine-tuning of existing methods will continue into the foreseeable future. Recent new examples include the immune algorithm, lightning flash algorithm, teaching–learning-based optimization algorithm, symbiotic organism search algorithm, enhanced firework algorithm, modified differential evolution, adaptive charged system search algorithm, distributed auction optimization algorithm, water cycle algorithm, and mine blast algorithm [62–87].

3.4 FORMULATION AND TESTING OF THE FFAE ALGORITHM

Meta-heuristic algorithms have performed very well within the optimization space. However, the existing methods have several drawbacks, as outlined in the above review.

Swarm-based algorithms have been shown to be superior to evolutionary techniques because they preserve the search space over subsequent iterations instead of evolutionary methods that eliminate aging data as soon as a new population is developed.

As such, the development of new robust nature-inspired, swarm-based, meta-heuristic and hybrid optimization methods will continue to grow in the foreseeable future as power systems become bigger and more complex. This work sought to develop and test a new method for the best location of FACTS devices named the filter feeding allogenic engineering (FFAE) algorithm [88, 89].

The results obtained were validated against the whale optimization algorithm and the fast voltage stability index (VSI).

3.4.1 Formulation of the FFAE Algorithm

Filter feeders are a group of ecosystem engineers that feed by squeezing oceanic matter and nutrient particles by passing the water over a filtration system. Some oceanic organisms that employ this feeding method include, but are not limited to, clams, krill, and sponges [90, 91].

Ecosystem engineers are oceanic organisms that impact the availability of nutrients to other organisms within the ecosystem by causing actual state changes in biotic or abiotic matter. They are broadly classified into two, namely, autogenic and allogenic engineers. Additionally, they impact the amount of nutrients available to other oceanic organisms. This can be achieved as a direct result of the structure they create (autogenic engineers) or by changing the biotic or abiotic forces (allogenic engineers).

Allogenic engineers move large amounts of material within the water through filter feeding as they move around, mainly due to nutrient stimuli within their environment. They affect the availability of nutrients to other organisms by changing biotic and abiotic forces through their body structure and biological activity. They can interfere with abiotic factors such as water residence time, hydrodynamic parameters, and water filtration, thus providing a residential habitat for other marine species.

The cells of allogenic engineers have their protoplasts covered by a thin and pliable layer. Membrane-bound proteins are involved in responding to given environmental stimuli by motion. They work as environmental monitors; e.g., if something in the water goes bad, the allogenic engineer filter feeders are the first to show the adverse effects.

The development of the FFAE algorithm was thus inspired by the feeding and motile behavior of allogenic engineers such as herring clams, sponges, baleen whales, krill, and ameboid protozoa.

The power network environment will be scanned in real time for set parameters to obtain their best solution in the same way the allogenic engineers respond to real-time stimuli in the oceanic environment.

The equation below shows the time rate of change of a given nutrient stimulus for the locomotion of allogenic engineers in seawater:

$$\frac{dC_m}{dt} = \left(k_m - \sum_{n=1}^{N} k_{mn} C_n - k_h\right) C_m + k_h C_{mo}, \tag{3.1}$$

where c_m = concentration of a given nutrient in seawater,
N = total number of constituents (dependent variables),
c_{mo} = the most influential chemical in the given water ecosystem, e.g., nitrogen,
t = time,
k_m = net production rate of the given nutrient minus natural decay, respiration, and sinking processes,
k_{mn} = rate coefficients for the uptake of c_m by other constituents c_n,
k_h = reciprocal of the hydrodynamic residence time.

The above equation takes care of a large number of coefficients, the majority of which are a function of c_n, light, temperature, and turbidity. It can be reduced to the prey–predator equations written as

$$\frac{dP}{dt} = k_p - \frac{F}{h} BP, \tag{3.2}$$

$$\frac{dB}{dt} = -k_b B + \alpha \beta_1 \beta_2 FPB, \tag{3.3}$$

where P = a given concentration,
B = filter feeder population,
k_p = given nutrient growth rate,
k_b = filter feeder mortality rate,
F = specific filtration rate for a given filter feeder,
h = depth of water,
α = conversion coefficient of the most influential chemical component, e.g., nitrogen, to the given type of nutrient,
β_1 = filter feeders' feeding efficiency,
β_2 = filter feeders' nutrient conversion coefficient.

Equation (3.2) shows that the rate of increase in time for a given nutrient equals the most influential chemical conversion efficiency minus the removal rate through filter feeding. The iterative determination of α, β_1, β_2, k_p, and k_b will be used to solve P and B, whereby P is the highest concentration that will act as the stimuli to attract the largest number of filter feeders (B).

The two main objectives during optimal placement of FACTS devices are the minimization of active power loss and voltage deviations using optimally placed shunt and/or series FACTS devices as formulated below.

The minimization of active power loss is formulated as follows:

$$Minimize\,F(x_1, x_2) = P_{loss} = \sum_{j=1}^{n}[G_i(V_i^2 + V_j^2 - 2V_iV_j cos\delta_{ij})]\,, \tag{3.4}$$

whereby

$$x_1 = [Q_{G1,\ldots\ldots\ldots,}Q_{GNPV,}V_{L1},\ldots,V_{LNPQ,}S_{L1},\ldots,S_{LNL}], \tag{3.5}$$

$$x_2 = [T_1,\ldots,T_{NT},\ldots V_{G1},\ldots,V_{GNPV}, Q_{C1},\ldots,Q_{CNC}]. \tag{3.6}$$

x_1 is the vector of dependent variables consisting of the reactive power output of the generators ($Q_{G1,\ldots\ldots\ldots,}Q_{GNPV}$), load voltages ($V_{L1},\ldots,V_{LNPQ}$), and transmission lines loading ($S_{L1},\ldots,S_{LNL}$).

x_2 is the vector of control variables consisting of transformer tap settings ($T_1,\ldots,T_{NT}$), generator voltages ($V_{G1},\ldots,V_{GNPV}$), and reactive power injections ($Q_{C1},\ldots,Q_{CNC}$).

In this case, G_i is the conductance of branch i; V_i and V_j are the magnitude of voltages in the sending bus and receiving bus, respectively; and δ_{ij} is the phase angle difference between the ith and the jth bus.

The equality constraints are represented by the power balance equations. The practical implication of this is to solve the load flow problem with active and reactive power constraints on the bus as formulated below.

$$P_{gi} - P_{di} - \sum_{j=1}^{87} V_iV_j(G_{ij}cos\theta_{ij} + B_{ij}sin\theta_{ij}) = 0, \tag{3.7}$$

$$Q_{gi} - Q_{di} - \sum_{j=1}^{87} V_iV_j\left(G_{ij}sin\theta_{ij} - B_{ij}cos\theta_{ij}\right) = 0, \tag{3.8}$$

where $i = 1,\ 2\ldots n$ = number of buses in the system = 87,

$\theta_{ij} = \theta_i - \theta_j$,

P_i, Q_i = injected active and reactive power in bus i,

P_{di}, Q_{di} = active and reactive power demand in bus i,

V_i, θ_i = bus voltage magnitude and angle on bus i,

G_{ij}, B_{ij} = conductance and susceptance of the element in the admittance matrix.

Inequality constraints reflect the limits on physical power system equipment and the limits in place to ensure power system security. They are the following:

$P_{gimin} \le P_{gi} \le P_{gimax}$ = lower and upper bounds on active power generation set at 10–90% of the machine rated value,

$Q_{gimin} \le Q_{gi} \le Q_{gimax}$ = lower and upper bounds on reactive power generation for each machine,

$t_{ijmin} \le t_{ij} \le t_{ijmax}$ = given lower and upper bounds on the transformer tap ratio,

$\alpha_{ijmin} \le \alpha_{ij} \le \alpha_{ijmax}$ = lower and upper bounds on phase shifting of variable transformers,

$|P_{ij}| \le P_{ijmax}$ = upper limit on the active power flow of line i–j,

$|S_i \leq S_i^{max}$ = transmission line loading capacity limit,

$V_{imin} \leq V_i \leq V_{imax}$ = lower and upper bounds on the bus voltage magnitude set at 0.9–1.05 per unit. (3.9) [92–102]

The minimization of voltage deviation is expressed as

$$VD = \sum_{i=1}^{n} \left| V_b - 1.0 \right|, \tag{3.10}$$

where n is the total number of buses in the transmission network and V_b is the bus voltage.

The proposed optimization algorithms are employed to solve the models given in Eqs. (3.1)–(3.10) to obtain the optimal location of FACTS devices for improving both the stability of the voltage and rotor angle system stability.

The FFAE algorithm for the optimal placement of FACTS devices was developed and executed as formulated above and by MATLAB's optimization toolbox guidelines. The algorithm executed was developed as per the flow in Figure 3.7.

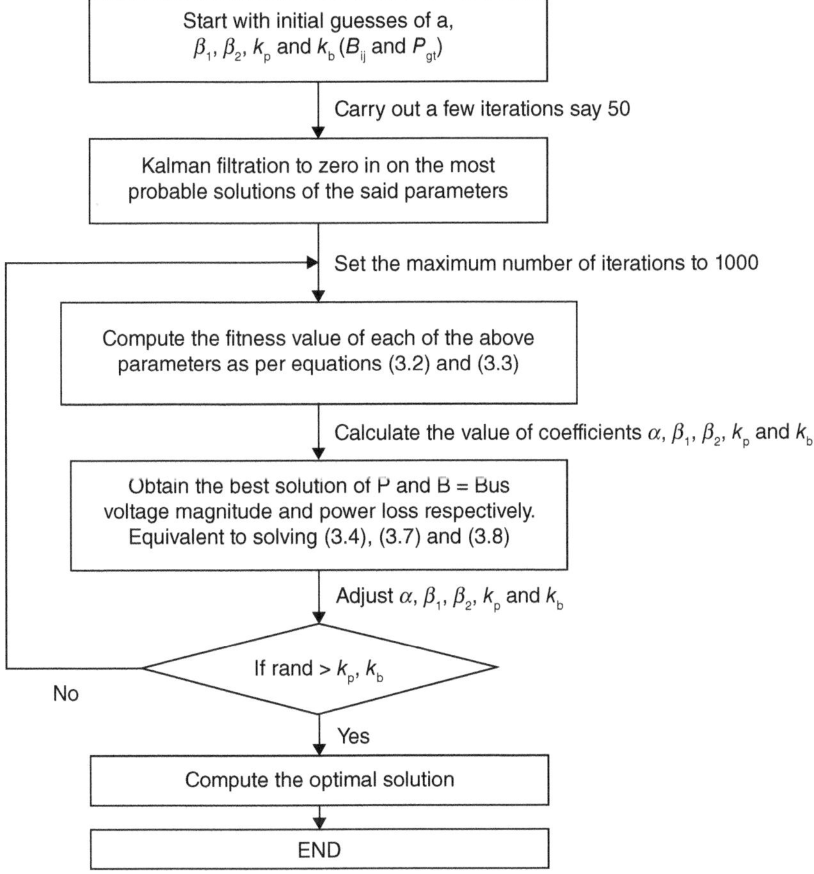

FIGURE 3.7 Flowchart of the FFAE Optimization Algorithm.

3.4.2 Formulation of the Whale Optimization Algorithm

This method was inspired by the hunting behavior of the humpback whale species. Humpback whales are part of a small group of mammals with spindle cells that handle rapid information processing and transfer. They can think, learn, judge, show emotions, and communicate in their language [45].

Humpback whales hunt for small groups of fish using a unique method called the bubble-net feeding technique. They swim in a synchronized shrinking circle, blowing bubbles and creating a loud and intense sound that ultimately forces fish upward as they herd and corner them together. The noise makes it very hard for the fish to escape from the trap. The lead whale dives deep from where it begins to blow air to create bubbles. The rest of the team members follows, building a shrinking fence around the fish prey. The whales then gather inside the forming bubble-net and rise to the surface with their mouths wide open to swallow the fish.

Bubble-net hunting has three distinct steps. The first one is to identify the location of the target prey and encircle them. The algorithm makes use of the current solution as the best fit. On the definition of the best search agents, the other agents will try to give an update of their positions by looking for the best search agent:

$$\vec{D}=\left|\vec{C}\cdot\vec{X}^{*}-\vec{X}(t)\right|, \tag{3.11}$$

$$X(t+1)=\vec{X}^{*}(t)-\vec{A}\cdot\vec{D}, \tag{3.12}$$

where t denotes the current iteration,

$A^{\rightarrow}$ and $C^{\rightarrow}$ are the coefficient vectors,

X^{*} is the position vector of the best solution obtained thus far,

$X^{\rightarrow}$ is the position vector,

$|\;|$ is the absolute value,

. is an element-by-element multiplication.

X^{*} should be updated if there is a better solution.

The vectors $A^{\rightarrow}$ and $C^{\rightarrow}$ are computed as follows:

$$\vec{A}=2\vec{a}\cdot\vec{r}-\vec{a}, \tag{3.13}$$

$$\vec{C}=2\vec{r}, \tag{3.14}$$

where $a^{\rightarrow}$ is reduced from 2 to 0 during the iteration period in both the exploration and exploitation phases and $r^{\rightarrow}$ is a random vector within the [0, 1] space.

The bubble-net technique ($|A|<1$) is formulated by using two methods. The first method is the shrinking circle technique, where the value of $a^{\rightarrow}$ is decreased in Equation (3.13). The variation range of $A^{\rightarrow}$ is also reduced by decreasing $a^{\rightarrow}$, i.e., $A^{\rightarrow}$ is a random value within $[-a, a]$, where a is reduced from 2 to 0 in the iteration period. The current position of the agent is updated to a new one between the actual position and the position of the current best agent by random guessing the values for $A^{\rightarrow}$ in the space [–1, 1].

The other approach is by using spiral updating. In this method, the distance between the whale and the prey located at (X, Y) and (X^*, Y^*), respectively, is computed first. A spiral equation is then created between the position of the whale and the prey to simulate the helix-shaped locomotion of humpback whales as follows:

$$\vec{X}(t+1) = \vec{D}\cdot e^{bl}\cos(\pi l) + \vec{X}^*(t), \tag{3.15}$$

where $D^{\rightarrow} = \left|X^{*\rightarrow}(t) - X^{\rightarrow}(t)\right|$ indicates the separation between the *i*th whale and the target prey, *b* defines the logarithmic curve of the spiral, *l* is an arbitrary number within [–1, 1], and . is n multiplication operation.

Humpback whales circle the prey by means of a reducing circle and along a spiral-shaped path at the same time. Each has an equal chance of being chosen by the whale at any given time, and the possibility is formulated as follows:

$$\vec{X}(t+1) = \vec{X}^*(t) - \vec{A}\cdot\vec{D} = \vec{D}\cdot e^{bl}\cos(2\pi l) + \vec{X}^*(t) \tag{3.16}$$

for $p < 0.5$ and $p < 0.5$, respectively, where *p* is a given number within [0,1].

In addition to the bubble-net method, the humpback whales perform a random search for prey using their positions relative to each other (exploration phase $\left|A^{\rightarrow}\right| \geq 1$). Here, $A^{\rightarrow}$ is used with the arbitrary values spread from +1 to –1 to trigger the agent to drift away from the reference point. Compared to the exploitation stage, the position of a search agent in the exploration stage is updated according to a randomly chosen search agent instead of the best search agent. This practice and $\left|A^{\rightarrow}\right| \geq 1$ stress the exploration and allow the whale optimization algorithm to do a global search, whose mathematical equations are as follows:

$$\vec{D} = \left|\vec{C}\cdot\vec{X}_{round} - \vec{X}\right|, \tag{3.17}$$

$$\vec{X}(t+1) = \vec{X}_{round} - \vec{A}\cdot\vec{D}, \tag{3.18}$$

where $X_{rand}^{\rightarrow}$ represents a random whale.

This is summarized in Figure 3.8.

3.4.3 The Fast Voltage Stability Index

The study of voltage stability is done using several methods that can be broadly grouped into dynamic and static methods. Dynamic methods use differential and algebraic equations, including generators, motors, and transformers in the power system. Static methods solve the differential equations obtained from power flow simulations of the network. They are done with given load increases until the voltage collapses, which allows the study of a wide range of system operating conditions [100, 101].

VSI is one of the sensitivity-based methods used to determine the optimal location of FACTS devices. This is done by tracking the power system output variables relative to a set threshold as the input variables change. The VSI methods are classified

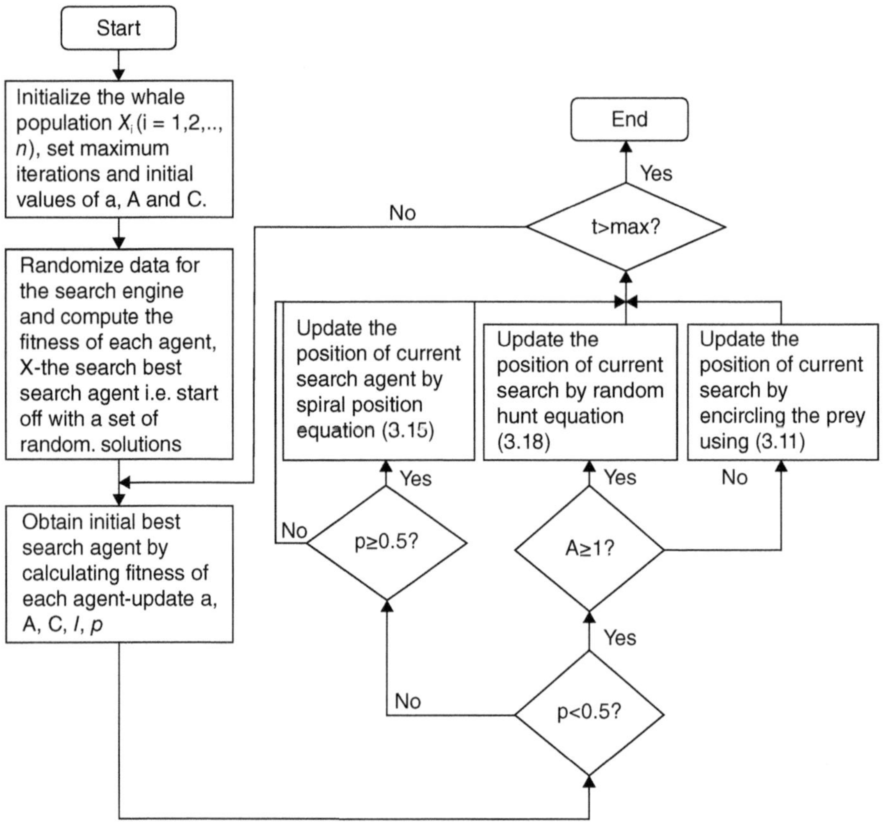

FIGURE 3.8 Whale Optimization Algorithm Flowchart.

into two, namely, the Jacobian matrix methods and the system variable techniques. Jacobian matrix-based VSI computes the voltage collapse point and hence determines the voltage stability margin. The VSI system variables use the admittance matrix and other variables such as bus voltages.

This research uses one of the reactive power adjustment methods called the fast voltage stability index (FVSI) because voltage stability and, by extension, loss reduction are key objectives of the study. The FVSI utilizes the admittance matrix, voltage, and reactive power measurements.

The two-bus model in Figure 3.9 is used to derive the FVSI.

This starts with obtaining the equation for the current flowing from bus i to bus j:

$$I_{ij} = \frac{u_i - u_j}{z_{ij}}. \tag{3.19}$$

i and j represent the sending and receiving ends, respectively, and bus i is taken as the reference bus with its voltage angle 0.

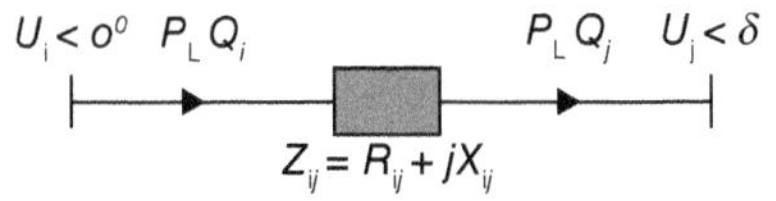

FIGURE 3.9 Two-Bus Line Model.

The apparent power on bus j is computed by multiplying (3.12) with the voltage at bus j as

$$I_{ij}=u_j\left(\frac{u_i-u_j}{z_{ij}}\right)^{*}=P_j+jQ_j. \tag{3.20}$$

The imaginary part of Equation (3.20) is the reactive power at bus j, which is given as

$$Q_j=\frac{u_iu_j(R_{ij}\sin\delta+X_{ij}\cos\delta-X_{ij}u_{ij}^2}{R_{ij}^2+X_{ij}^2}. \tag{3.21}$$

This can be rewritten as u_j:

$$u_j^2-u_ju_i\left(\frac{R_{ij}}{X_{ij}}\sin\delta+\cos\delta+X_{ij}+\frac{R_{ij}^2}{X_{ij}}\right)Q_j. \tag{3.22}$$

The power system is stable only for real solutions of Equation (3.22), i.e.,

$$\left[\left(\frac{R_{ij}}{X_{ij}}\sin\delta+\cos\delta\right)u_i\right]^2-4\left(X_{ij}+\frac{R_{ij}^2}{X_{ij}}\right)Q_j\geq 0. \tag{3.23}$$

Simplifying Equation (3.23) above gives
($\delta\approx 0,\ R_{ij}\sin\delta\approx 0$ and $X_{ij}\cos\delta\approx X_{ij}$), which gives

$$\left(X_{ij}u_i\right)^2\geq\left(4X_{ij}+R_{ij}^2\right)Q_j. \tag{3.24}$$

The FVSI is then defined as

$$\text{FVSI}=\frac{4Z_{ij}^2Q_j}{u_i^jX_{ij}}\leq 1. \tag{3.25}$$

As seen in Equation (3.25), stability is maintained as long as FVSI ≤ 1.

3.5 TESTING OF THE FFAE ALGORITHM

3.5.1 FFAE Algorithm for the Optimal Placement of FACTS Devices

The FFAE algorithm was then tested for the best location of the FACTS devices to improve and control the stability of the power system. Kenya's 87-bus, 25-generator,

132-KV, and 220-KV transmission network was used as the test case in MATLAB's environment. The two variables of bus voltage magnitude and active power loss were solved iteratively using the proposed FFAE. A meta-heuristic technique whale optimization algorithm (WOA) and a sensitivity-based method (FVSI) were used to validate the accuracy of the new algorithm.

The two criteria used for the best location of the FACTS devices were the minimization of the active power loss and voltage deviations.

Minimizing the active power loss requires that

$$F(x_1,x_2)=P_{\text{loss}}=\sum_{j=1}^{n}\left[G_i(V_i^2+V_j^2-2V_iV_j\cos\delta_{ij})\right] \quad (3.26)$$

where

$$x_1 = \left[Q_{G1},\cdots,Q_{\text{GNPV}},V_{L1},\cdots,V_{\text{LNPQ}},S_{L1},\cdots,S_{\text{LNL}}\right] \quad (3.27)$$

$$x_1 = \left[Q_{G1},\cdots,Q_{\text{GNPV}},V_{L1},\cdots,V_{\text{LNPQ}},S_{L1},\cdots,S_{\text{LNL}}\right] \quad (3.28)$$

x_1 is the vector of dependent variables made up of the reactive power output of the generators

$(Q_{G1,\dots\dots\dots,}Q_{GNPV})$, load voltages $(V_{L1},\dots..,V_{LNPQ})$, and transmission line loading $(S_{L1},\dots,S_{LNL})$.

x_2 is the vector of control variables made up of transformer tap settings $(T_1,\dots,T_{NT})$, generator voltages $((V_{G1},\dots,V_{GNPV})$, and reactive power injections $(Q_{C1},\dots,Q_{CNC})$.

In this case, G_i is the conductance of i; V_i and V_j are the magnitude of voltages at the sending bus and receiving bus, respectively; and δ_{ij} is the angle difference between ith and jth buses.

The power balance equations represent the equality constraints. The practical implication of this is to solve the load flow problem with active and reactive power constraints on the bus as follows:

$$P_{gi}-P_{di}-\sum_{j=1}^{87}V_iV_j(G_{ij}\cos\theta_{ij}+B_{ij}\sin\theta_{ij})=0 \quad (3.29)$$

$$Q_{gi}-Q_{di}-\sum_{j=1}^{87}V_iV_j(G_{ij}\sin\theta_{ij}-B_{ij}\cos\theta_{ij})=0 \quad (3.30)$$

where $i=1,\ 2\dots n$ = number of buses in the system = 87,

$\theta_{ij}=\theta_i-\theta_j$,

P_i,Q_i = injected active and reactive power in bus i,

P_{di},Q_{di} = active and reactive power demand in bus i,

V_i,θ_i = bus voltage magnitude and angle on bus i,

G_{ij}, B_{ij} = conductance and susceptance of the element in the admittance matrix.

Inequality constraints enforce the power system and ensure there is a power balance in the system. The equations are detailed in the following.

$P_{gimin} \le P_{gi} \le P_{gimax}$ = lower and upper bounds on active power generation set at 10–90% of the machine rated value,

$Q_{gimin} \le Q_{gi} \le Q_{gimax}$ = lower and upper bounds on reactive power generation for each machine,

$t_{ijmin} \le t_{ij} \le t_{ijmax}$ = given lower and upper bounds on the transformer tap ratio,

$\alpha_{ijmin} \le \alpha_{ij} \le \alpha_{ijmax}$ = lower and upper bounds on phase-shifting of variable transformers,

$|P_{ij}| \le P_{ijmax}$ = upper limit on the active power flow of line *i–j*,

$| S_i \le S_i^{max}$ = transmission line loading capacity limit,

$V_{imin} \le V_i \le V_{imax}$ = lower and upper bounds on the bus voltage magnitude set at 0.9–1.05 per unit (3.31) [102–111].

The reduction of voltage deviation to the minimum is expressed as

$$\text{VD} = \sum_{i=1}^{n} |V_b - 1.0| \tag{3.32}$$

where n is the total number of buses in the transmission network and V_b is the bus voltage.

The optimization algorithm was used to solve models given in Equations (3.27)–(3.32) to obtain the best location for FACTS devices to improve both the voltage and rotor angle system stability.

Equations (3.2) and (3.3) show that the increase in time of a given nutrient equals the most influential chemical conversion efficiency minus the removal rate through filter feeding. The iterative determination of α, β_1, β_2, k_p, and k_b is used to solve for P and B. The highest concentration (P) will act as stimuli to attract the largest number of filter feeders (B), which will be our optimal solution equivalent to the given generator outputs and power losses obtained by solving Equations (3.29) and (3.30).

The load and generator data with set power output constraints will be equated to given parameters that affect the nutrient concentration in the seawater. These are the stimuli that normally result in the movement of most allogenic species toward the direction with the highest nutrient concentration in the sea. The optimization will start with some initial solution guesses. After a few iterations, Kalman filtration will be applied to zero in on the most probable solutions, which will be iterated until the end. The filtration process will be used to predict which of the initial guesses has higher chances of convergence by walking the entire probability distribution of each after a few iterations. This enables mathematical distinction between phenomena and noumena through complete statistical characterization of the estimation problem at hand.

The obtained results were compared with an existing meta-heuristic placement method (whale optimization algorithm) and a sensitivity-based optimal placement method (FVSI).

The best location of FACTS devices started with the extraction of Kenya's transmission network data and the applicable equality and inequality constraints, as shown in Tables A.3–A.5. This was obtained from the Kenya Power SCADA system. The network was modeled in MATLAB's Simulink, as shown in Figure A2. Two key variables of the operation of the power system, namely, the reduction of active power losses and maintaining voltage profiles within the acceptable limits of 0.90–1.05 per unit, were considered for the optimal placement of FACTS devices. Kenya's recorded peak demand of 1960 MW was used.

The FFAE algorithm already developed and tested on the economic dispatch problem was used for the placement of the device. Proper adherence to MATLAB's optimization toolbox guidelines was done. The execution was as shown in Figure 3.7 [102].

Active power loss Equation (3.26) and equality constraint Equations (3.31) and (3.32) were formulated in the Kenyan transmission network and solved using the new optimization algorithm. P and B parameters were solved iteratively to give the various transmission bus voltage and power losses, respectively.

Kalman filtration was used to speed up the iteration process by focusing on the most probable solutions with a high probability of convergence [104].

The whale optimization algorithm was then used to test the accuracy of the FFAE algorithm. This was done as per the flowchart in Figure 3.8 and using the guidelines and applicable source code in MATLAB. The simulation done was as shown in Figure 3.10 [109, 110].

Continuous iterations were carried out during which the parameters a, A, C, and the system Y-bus matrix were continuously updated. A comparison of the value of the fitness solution with the search agent best fit was made for each iteration. The iterations were terminated at the maximum number of 1,000 iterations. The fitness values obtained were treated as the best solution for the best location of the FACTS devices. Parameters A and C represent the voltage magnitude and active power loss, respectively.

Finally, FVSI was used to test the accuracy of FFAE. A load flow solution was carried out on Kenya's 87-bus transmission network using the MATLAB PSAT platform and data in Tables A.3–A.5. The upper voltage limit was capped at 1.05 per unit. The load flow solution gave the magnitude and phase angle of the voltage at every bus and the power flowing in each transmission line. These data were the basis of the FVSI computation for the optimal placement of FACTS devices.

The FVSIs were computed by slowly increasing the reactive power at a given load bus until the load flow stops converging while maintaining the real power constant at the said load bus.

At the tipping point, the load on a given bus was determined to be the maximum loading point for the said bus. This was sorted in ascending order, with the smallest

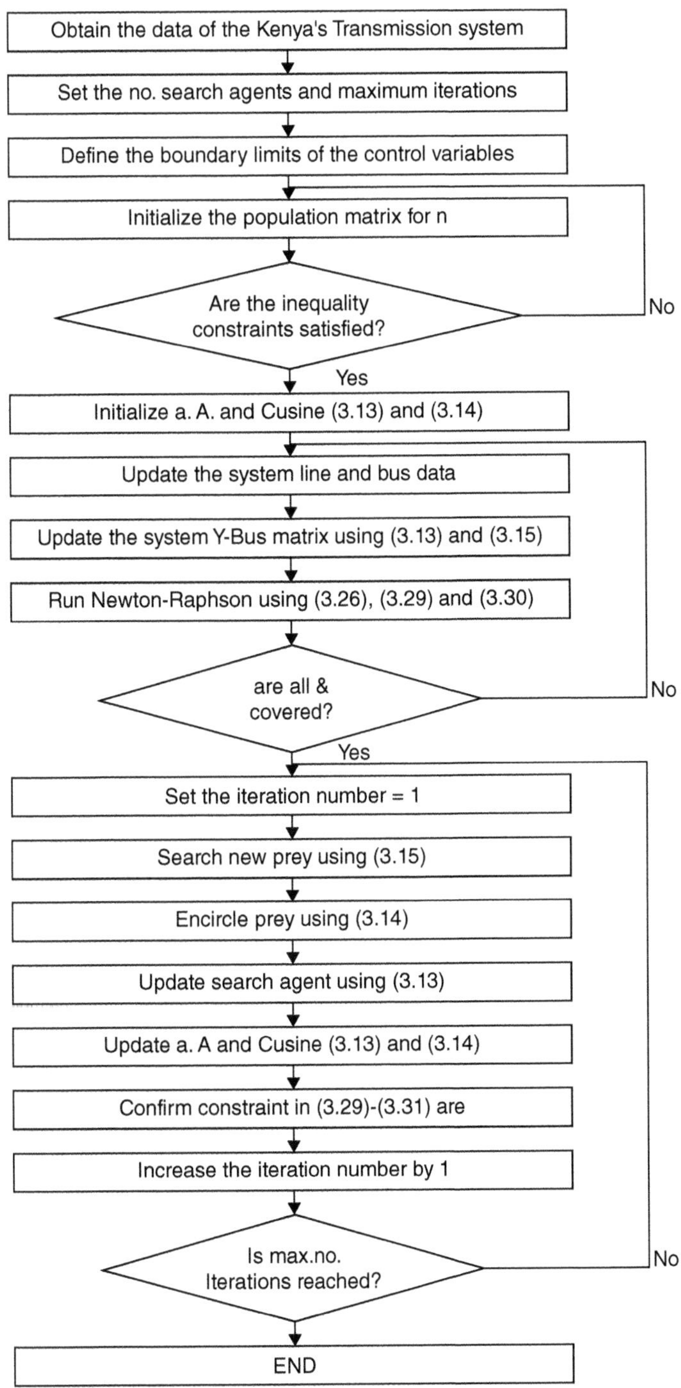

FIGURE 3.10 Flowchart of the Whale Optimization Algorithm in the Kenyan Transmission Network.

value being the highest ranked for all the load buses using Equation (3.25). In this case, the bus was the weakest bus and the best location for FACTS devices.

3.6 RESULTS AND DISCUSSION ON THE FFAE ALGORITHM

3.6.1 Optimal Placement of FACTS Devices Using the FFAE Algorithm

The FFAE algorithm was used to solve the voltage magnitude (*P*) and active power loss (*B*) at the various transmission buses. For the whale optimization algorithm, parameters *A* and *C* representing the magnitude of the transmission bus voltage and power loss, respectively, were computed.

The computation time was 0.283 seconds and 0.308 seconds for FFAE and the whale optimization algorithm, respectively. The FFAE algorithm performed better, mainly because Kalman filtration was used to achieve targeted iterations and formulated it as a more compressed algorithm. The FFAE algorithm summed the total active power losses to 99.911%, while the whale optimization algorithm gave a total of 99.770%, indicating the superior computational ability of the new method.

The buses were ranked from the weakest to the strongest, with the first 30 buses tabulated. For the purposes of this work, the rest of the buses had minimal losses and were not ranked.

The FVSI was used to determine the highest loading capacity of the various load buses, and the same was also ranked.

The results obtained for the three placement techniques of FFAE, WOA, and FVSI are shown in Table 3.1.

The total active power loss was 202.56 MW or 10.33% of the total system peak load demand of 1960 MW.

The convergence characteristics of the algorithms are as shown in Table 3.2 and Figure 3.11.

Figure 3.12 shows the active power loss as a percentage of the total system loss for the 87-bus Kenyan system.

The eight weakest buses in Kenya's transmission network account for 55% of total system losses and have the highest bus voltage deviations, thus forming the best candidates for the placement of FACTS devices. The eight buses said are listed in Table 3.3, starting with the weakest (Kisumu 132 KV) to the strongest (Muhoroni 132 KV).

In the FFAE algorithm, the buses with the highest amount of active power loss and the largest percentage of voltage deviation are determined to be the weakest buses, as seen in Tables 3.1 and 3.3. This was done by iterative and simultaneous solution of Equations (3.2) and (3.3).

The Kisumu 132 KV and Lessons 132 KV buses are the weakest buses in the Kenyan transmission network, and thus the best candidates for the placement of FACTS devices. Both are located at the leading load centers of the western part of the country. This is mainly due to the long distance from the geothermal power generation hub (in the central part of the country). The western part of the country has very low generation levels compared to the load demand. The primary role of the Muhoroni 28 MW peak gas plant has been to generate reactive power during peak systems, but it

TABLE 3.1
Voltage Profiles, Active Power Losses, and Loadability Limits

	FFAE				WOA				FVSI		
Bus	Bus Voltage (KV)	% Voltage Deviation	Active Power Loss as % of Total System Load	Rank	Bus Voltage (KV)	% Voltage Deviation	Active Power Loss as % of Total System Load	Rank	FVSI	Loadability (MVAr)	Rank
Kindaruma 132.00	131.16	0.636363636	0.345576619		131.19	0.613636364	0.34063981				
Gitaru 132.00	131.22	0.590909091	0.320892575		131.33	0.507575758	0.315955766				
Githambo 132.00	126.23	4.371212121	2.468404423	12	126.25	4.356060606	2.458530806	12	0.9039	35.7	12
Kamburu 132.00	131.26	0.560606061	0.34063981		131.26	0.560606061	0.335703002				
Garissa 132.00	131.24	0.575757576	0.370260664		131.24	0.575757576	0.365323855				
Meru 132.00	131.26	0.560606061	0.315955766		131.32	0.515151515	0.315955766				
Kutus 132.00	131.44	0.424242424	0.38507109		131.44	0.424242424	0.380134281				
Masinga 132.00	131.27	0.553030303	0.390007899		131.56	0.333333333	0.38507109				
Olkaria I 132.00	131.32	0.515151515	0.404818325		131.32	0.515151515	0.399881517				
Athi river 220.00	209.23	4.895454545	3.949447077	7	209.21	4.904545455	3.93957346	7	0.9298	27.6	7
Wote 132.00	131.14	0.651515152	0.345576619		131.14	0.651515152	0.345576619				
KDP3 132.00	131.88	0.090909091	0.202409163		131.21	0.598484848	0.202409163				
Tiomin 132.00	131.32	0.515151515	0.375197472		131.32	0.515151515	0.375197472				
isinya 220.00	218.58	0.645454545	0.25671406		218.58	0.645454545	0.25671406				
Vipr 132.00	131.42	0.439393939	0.232030016		131.42	0.439393939	0.232030016				
Mangu 132.00	126.99	3.795454545	2.221563981	13	127.01	3.78030303	2.196879937	13	0.9021	36.5	13
Menengai 132.00	131.24	0.575757576	0.25671406		131.24	0.575757576	0.261650869				
Msac 132.00	131.58	0.318181818	0.38507109		131.58	0.318181818	0.390007899				
Sarm 132.00	131.52	0.363636364	0.320892575		131.28	0.545454545	0.320892575				

(Continued)

TABLE 3.1 *(Continued)*
Voltage Profiles, Active Power Losses, and Loadability Limits

Bus	FFAE				WOA				FVSI		
	Bus Voltage (KV)	% Voltage Deviation	Active Power Loss as % of Total System Load	Rank	Bus Voltage (KV)	% Voltage Deviation	Active Power Loss as % of Total System Load	Rank	FVSI	Loadability (MVAr)	Rank
Olkau2 132.00	131.33	0.507575758	0.261650869		131.33	0.507575758	0.266587678				
Olkau 220.00	218.98	0.463636364	0.30114534		218.98	0.463636364	0.30114534				
Olkaria 4 132.00	131.18	0.621212121	0.38507109		131.28	0.545454545	0.38507109				
Oau 132.00	131.19	0.613636364	0.38507109		131.19	0.613636364	0.38507109				
Tsavo 132.00	131.26	0.560606061	0.320892575		131.26	0.560606061	0.320892575				
Eldoret 132.00	129.12	2.181818182	0.888625592	20	129.14	2.166666667	0.888625592	20	0.8865	52.4	20
Muhoroni 132.00	125.79	4.704545455	3.455766193	8	125.81	4.689393939	3.455766193	8	0.9233	29.4	8
Isiolo 132.00	131.28	0.545454545	0.25671406		131.28	0.545454545	0.25671406				
Kisumu 132.00	124.21	5.901515152	11.35466035	1	124.01	6.053030303	11.34478673	1	0.9523	21.6	1
Webuye 132.00	125.98	4.560606061	2.962085308	10	125.96	4.575757576	2.962085308	10	0.9219	31.2	9
Kiganjo 132.00	131.17	0.628787879	0.320892575		131.17	0.628787879	0.320892575				
Kilifi 132.00	125.92	4.606060606	3.20892575	9	125.92	4.606060606	3.20892575	9	0.9121	33.1	11
Kato 132.00	131.14	0.651515152	0.370260664		131.14	0.651515152	0.370260664				
Narok 132.00	131.17	0.628787879	0.276461295		131.17	0.628787879	0.276461295				
Musaga 132.00	126.05	4.507575758	2.715244866	11	126.08	4.484848485	2.715244866	11	0.9204	32.7	10
Lessos 132.00	124.88	5.393939394	8.886255924	2	124.84	5.424242424	8.876382306	2	0.9503	22.3	2
Tororo 132.00	131.26	0.560606061	0.12835703		131.26	0.560606061	0.12835703				
Lanet 132.00	129.38	1.984848485	0.789889415	21	129.35	2.007575758	0.789889415	21	0.8833	54.1	22

TABLE 3.1 *(Continued)*
Voltage Profiles, Active Power Losses, and Loadability Limits

	FFAE				WOA				FVSI		
Bus	Bus Voltage (KV)	% Voltage Deviation	Active Power Loss as % of Total System Load	Rank	Bus Voltage (KV)	% Voltage Deviation	Active Power Loss as % of Total System Load	Rank	FVSI	Loadability (MVAr)	Rank
Kitale 132.00	129.84	1.636363636	0.691153239	22	129.84	1.636363636	0.691153239	22	0.8792	55.6	23
Loiya 132.00	131.28	0.545454545	0.355450237		131.28	0.545454545	0.355450237				
Naivasha 132.00	131.33	0.507575758	0.370260664		131.33	0.507575758	0.370260664				
Sultan 132.00	131.19	0.613636364	0.350513428		131.19	0.613636364	0.350513428				
Kiboko 132.00	131.48	0.393939394	0.30114534		131.48	0.393939394	0.30114534				
Voi 132.00	131.62	0.287878788	0.335703002		131.62	0.287878788	0.335703002				
Soilo 132.00	131.65	0.265151515	0.370260664		131.65	0.265151515	0.370260664				
Makutano 132.00	129.96	1.545454545	0.592417062	23	129.98	1.53030303	0.592417062	23	0.8843	53.2	21
Ruaraka 132.00	127.96	3.060606061	1.481042654	16	127.98	3.045454545	1.481042654	16	0.8972	40.3	16
Nrbn 220.00	208.55	5.204545455	7.898894155	3	208.48	5.236363636	7.884083728	3	0.9478	22.9	3
Mumias 132.00	128.64	2.545454545	1.234202212	17	128.66	2.53030303	1.234202212	17	0.8924	45.9	18
Sondu 132.00	131.18	0.621212121	0.355450237		131.18	0.621212121	0.355450237				
Kieni 132.00	131.26	0.560606061	0.325829384		131.26	0.560606061	0.325829384				
Kamburu 220.00	218.88	0.509090909	0.335703002		218.88	0.509090909	0.335703002				
mwingi 132.00	131.28	0.545454545	0.34063981		131.28	0.545454545	0.34063981				
Athi 220.00	214.87	2.331818182	1.184834123	18	214.87	2.331818182	1.160150079	18	0.8969	42.8	17
Kiambere 132.00	131.18	0.621212121	0.306082148		131.18	0.621212121	0.306082148				
Rangala 132.00	131.42	0.439393939	0.370260664		131.42	0.439393939	0.370260664				

(Continued)

TABLE 3.1 *(Continued)*
Voltage Profiles, Active Power Losses, and Loadability Limits

	FFAE				WOA				FVSI		
Bus	Bus Voltage (KV)	% Voltage Deviation	Active Power Loss as % of Total System Load	Rank	Bus Voltage (KV)	% Voltage Deviation	Active Power Loss as % of Total System Load	Rank	FVSI	Loadability (MVAr)	Rank
Turkwel 132.00	131.29	0.537878788	0.320892575		131.29	0.537878788	0.320892575				
Olkaria II 132.00	131.19	0.613636364	0.370260664		131.19	0.613636364	0.370260664				
Sangoro 132.00	131.42	0.439393939	0.370260664		131.42	0.439393939	0.370260664				
Embakasi 220.00	130.21	1.356060606	0.493680885	24	130.21	1.356060606	0.493680885	24	0.8745	58.7	25
Ishiara 132.00	131.31	0.522727273	0.335703002		131.31	0.522727273	0.335703002				
Dandora 220.00	217.55	1.113636364	0.46899684	25	217.59	1.095454545	0.46899684	25	0.8769	57.1	24
Kakuyuni 220.00	218.56	0.654545455	0.335703002		218.56	0.654545455	0.335703002				
Rabai 220.00	218.66	0.609090909	0.350513428		218.66	0.609090909	0.350513428				
Garsen 220.00	218.84	0.527272727	0.103672986		218.84	0.527272727	0.103672986				
Suswa 220.00	218.86	0.518181818	0.133293839		218.86	0.518181818	0.133293839				
City 220.00	217.85	0.977272727	0.419628752	26	217.85	0.977272727	0.419628752	26	0.8734	60.2	26
Gaso 132.00	131.33	0.507575758	0.375197472		131.33	0.507575758	0.375197472				
Kwale 132.00	131.18	0.621212121	0.360387046		131.18	0.621212121	0.360387046				
Juja rd 132.00	125.21	5.143939394	6.911532385	4	125.18	5.166666667	6.911532385	4	0.9421	23.8	4
Lessos 220.00	208.65	5.159090909	5.924170616	5	208.64	5.163636364	5.916765403	5	0.9388	24.9	5
Maungu 132.00	131.25	0.568181818	0.306082148		131.25	0.568181818	0.306082148				
Rabai 132.00	131.2	0.606060606	0.315955766		131.2	0.606060606	0.315955766				
KDP1 132.00	131.56	0.333333333	0.335703002		131.56	0.333333333	0.335703002				

TABLE 3.1 *(Continued)*
Voltage Profiles, Active Power Losses, and Loadability Limits

	FFAE				WOA				FVSI		
Bus	**Bus Voltage (KV)**	**% Voltage Deviation**	**Active Power Loss as % of Total System Load**	**Rank**	**Bus Voltage (KV)**	**% Voltage Deviation**	**Active Power Loss as % of Total System Load**	**Rank**	**FVSI**	**Loadability (MVAr)**	**Rank**
Jomvu 132.00	131.42	0.439393939	0.390007899		131.42	0.439393939	0.390007899				
Nbam 132.00	131.56	0.333333333	0.365323855		131.56	0.333333333	0.365323855				
RBPL 132.00	131.88	0.090909091	0.34063981		131.88	0.090909091	0.34063981				
Konza 132.00	131.58	0.318181818	0.157977883		131.58	0.318181818	0.157977883				
Lamu 220.00	218.21	0.813636364	0.414691943	27	218.23	0.804545455	0.414691943	27	0.8722	61.2	27
Orpower 132.00	131.27	0.553030303	0.370260664		131.27	0.553030303	0.370260664				
Nbnorth 220.00	209.08	4.963636364	4.936808847	6	209.19	4.913636364	4.926935229	6	0.9306	26.2	6
Nanyuki 132.00	130.97	0.78030303	0.409755134	28	130.96	0.787878788	0.409755134	28	0.8707	63.4	28
Ndhiwa 132.00	127.85	3.143939394	1.727883096	15	127.86	3.136363636	1.727883096	15	0.9014	37.8	14
Awendo 132.00	128.95	2.310606061	1.086097946	19	128.94	2.318181818	1.086097946	19	0.8873	48.8	19
Kega 132.00	127.26	3.590909091	1.974723539	14	127.29	3.568181818	1.974723539	14	0.9004	38.9	15
Sotik 132.00	131.02	0.742424242	0.404818325	29	131.02	0.742424242	0.404818325	29	0.8693	65.7	29
Bomet 132.00	131.13	0.659090909	0.394944708	30	131.09	0.689393939	0.394944708	30	0.8594	67.2	30
Csit 132.00	131.25	0.568181818	0.350513428		131.24	0.575757576	0.350513428				
			99.91113744				**99.7703839**				

TABLE 3.2
Convergence Characteristics of FFAE Algorithm and WOA

Algorithm	Computation Time (s)	Number of Iterations
FFAE	0.283	776
WOA	0.308	852

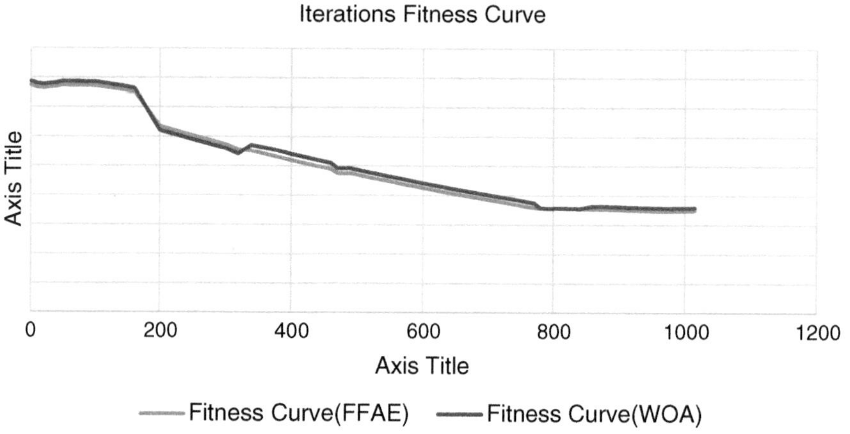

FIGURE 3.11 Convergence Curve of the WOA and FFAE Algorithm.

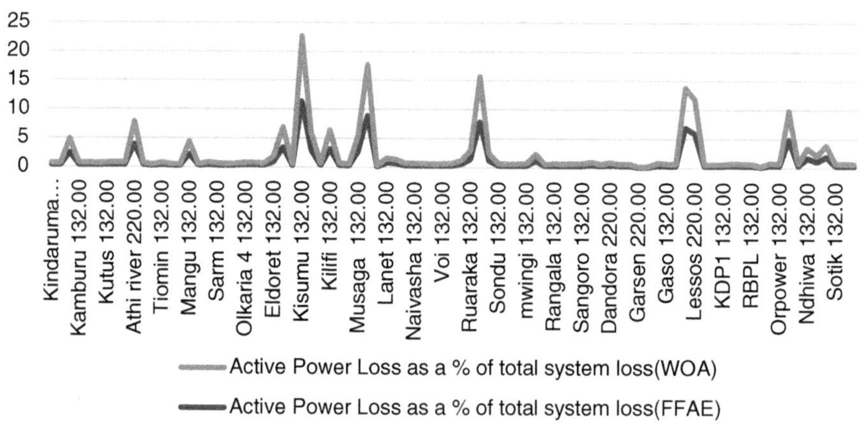

FIGURE 3.12 Active Power Loss as a Percentage of the Total Loss of the System for the Kenyan Transmission Network.

TABLE 3.3
Weakest Buses for the Optimal Location of FACTS Devices

Bus	Rank
Kisumu 132.00	1
Lessos 132.00	2
Nrbn 220.00	3
Juja Road 132.00	4
Lessos 220.00	5
Nairobi North 220.00	6
Athi River 220.00	7
Muhoroni 132.00	8

has not matched the heavy energy demand of the west Kenya load center of Kisumu city and its environs.

The next two candidates are in the Nairobi-Juja 132 KVA and Nrbn 220 KV buses. This is due to the high load in the capital city, which stands at 58% of the total national demand. The main challenge is the constrained energy transmission capacity from the coastal, Naivasha, and Mount Kenya generation centers to Nairobi at system peak, usually recorded around 8 pm. The Iberafrica diesel power plant in Nairobi was installed to support the voltage during the peak of the system.

A large number of buses with very low voltage deviations are the buses at the power generation sites with minimal load in the vicinity. These include Gitaru, Masinga, Olkaria, and others, as seen in Table 3.1.

The FFAE algorithm had a faster computation time of 0.283 seconds compared to 0.308 seconds by the WOA. This is mainly due to the use of Kalman filtration to achieve targeted iterations for a more compressed algorithm. The algorithm also gave more accurate results compared to the whale optimization algorithm and the FVSI. The sum of total active power losses was 99.911% and 99.770% for the filter-feeding optimization algorithm and the whale optimization algorithm, respectively, pointing to the superior computational ability of the new algorithm. This represents 202.38 MW and 202.095 MW for FFAE and WOA, respectively, against a total system loss of 202.56 MW, indicating the superior computational ability of the new algorithm. Figure 3.11 further illustrates the better computational efficiency in terms of losses for each of the 87 buses in the Kenyan transmission network.

It is good to note that the two meta-heuristic techniques gave similar results in terms of the ranking of the buses. The results of the sensitivity-based method differed slightly for 11 buses ranked number nine onward, as seen in Table 3.1. This can be attributed to the degraded computational efficiency of sensitivity-based methods when dealing with nonlinear, multivariable, and constrained real-world power systems. FVSI is manual and prone to human errors, and tends to do a simple estimate of the solution according to Equation (3.25). As power systems become bigger, multivariable, highly nonlinear, more complex, and interconnected, sensitivity-based methods tend to become inaccurate, as explained earlier.

3.7 FUTURE OUTLOOK OF POWER SYSTEM OPTIMIZATION

Optimal utilization of the existing network is critical in addressing the capital and environmental challenges of expanding power networks. Thus, the development of a new swarm-based FACTS devices optimal placement technique that hinges on the accurate summation of active power loss and voltage stability rhymes well with the agenda of enhanced power infrastructure utilization.

Due to the limitations and drawbacks of the existing meta-heuristic algorithms, improvements of the existing algorithms and developments of new ones will continue in the near future. The FFAE algorithm developed and tested in this work is a good example.

The use of big data analytics and the emerging and more refined artificial intelligence tools will go a long way in helping improve the accuracy of the existing meta-heuristic optimization tools.

Upcoming technologies as exemplified in the aggregators, virtual power plants, and transmission lines, and blockchain business will play a major role in the development of new power system optimization methods.

The new FFAE algorithm can be improved by refining the predator–prey equations as marine biologists continue to study and understand ecosystem engineers better. Most of the existing meta-heuristic optimization methods have their drawbacks. Thus, one area of improvement is combining the new algorithm with an existing method to take advantage of the strong traits of each.

Deeper penetration of intermittent renewable energy also necessitates serious research into the convergence of storage technologies and deployment of FACTS devices with respect to power system optimization.

As power systems become bigger, more stressed, interconnected, and complex, there is a need for increased research on the hybrid switched FACTS devices such as the Static Synchronous Compensator (STATCOM) with storage, Fault Current Limiters (FCL), and Thyristor Controlled Voltage Limiter (TCVL).

3.8 CHAPTER SUMMARY

The chapter began with the formulation of the proposed FFAE algorithm. This was done alongside the whale optimization algorithm and the FVSI, all of which were used to test and validate the new algorithm.

The FFAE algorithm was used for the best placement of FACTS devices in the Kenyan transmission system against WOA and the FVSI. The FFAE algorithm had a faster computation time of 0.283 seconds compared to 0.308 seconds by the WOA. Total active power losses were 99.911% and 99.770% for the filter feeding optimization algorithm and the whale optimization algorithm, respectively, indicating that the new algorithm had better accuracy.

From the above, it is evident that the new FFAE algorithm performed much better compared to the existing meta-heuristic algorithms. This shows the efficiency of both the computation methods and the accuracy of the new optimization algorithm.

REFERENCES

[1] N. Singh and P. Agnihotri, "Power system stability improvement using FACTS devices," in *International Journal of Advanced Research and Development*, vol. 3(4), pp. 170–176, 2018.

[2] S. Kumar et al., "Stability improvement of multimachine power system network using STATCOM and UPFC," in *International Journal for Research in Applied Science and Engineering Technology*, vol. 7, pp. 179–188, 2019.

[3] K. Rupika and K.K. Sharma, "Voltage regulation using FACTS devices: A review," in *International Journal of Pure and Applied Mathematics*, vol. 119(16), pp. 2207–2214, 2018.

[4] G. Pavani et al., "Power system stability improvement using FACTS devices," in *Airo International Research Journal*, vol. 7, pp. 1–11, 2016.

[5] S.D. Nascimento and M. Gouvea, "Voltage stability enhancement in power systems with automatic FACTS device allocation," in *Energy Procedia*, vol. 107, pp. 60–67, 2017.

[6] M.R. Al-Masud et al., "Capacity enhancement and voltage stability improvement of power transmission line by series compensation," in *EAI-Endorsed Transactions on Energy Web*, pp. 1–10, 2019.

[7] P. Tripathi and G.P. Pandiya, "Survey in the impact of FACTS controllers on power system performance," in *International Journal of Engineering Trends and Technology*, vol. 46(1), pp. 24–28, 2017.

[8] B.S. Kumar, "Loadability enhancement with FACTS devices using gravitational search algorithm," in *International Journal of Electrical Power and Energy Systems*, vol. 3(4), pp. 470–479, 2016.

[9] B. Bhattacharyya and S. Kumar, "Approach for the solution of transmission congestion with multiple-type FACTS devices," in *IET Generation, Transmission and Distribution*, vol. 10, pp. 2802–2809, 2016.

[10] A.M. Blumsack, "Transfer capability improvement through market-based operation of series FACTS devices," in *IEEE Transactions on Power Systems*, vol. 31(5), pp. 3702–3714, 2018.

[11] H. Liao and J.V. Milanovi, "On capability of different FACTS devices to mitigate a range of power quality phenomena," in *IET Generation, Transmission and Distribution*, vol. 11(5), pp. 1202–1211, 2017.

[12] A. Majeed et al., "Power system stability, efficiency and controllability improvement using FACTS devices," in *International Journal of Engineering Science and Computing*, vol. 9(1), pp. 19590–19596, 2019.

[13] B. Singh and R. Kumar, "A comprehensive survey on enhancement of system perfomances by using different types of FATCS controller in power systems with static and realistic load models," in *Energy Reports*, Elsevier, vol. 6, pp. 55–79, 2019.

[14] M.M. Khan et al., "Stability enhancement in multimachine power system by FACTS controller," in *International Conference on Global Trends in Signal Processing, Information Computing, and Communication*, pp. 414–418, 2016.

[15] X.P. Zhang et al., "FACTS devices and applications," in *Flexible AC Transmission Systems: Modelling and Control*, Springer, pp. 1–26, 2006.

[16] K.D. Dhaked and M. Lalwani, "A comprehensive review on a D-FACTS controller: Enhanced power flow controller (EPFC)," in *International Journal Advances in Engineering and Technology*, vol. 10(1), pp. 84–92, 2017.

[17] P.B. Mali and N.S. Mahajan, "Improving the voltage stability and performance of FACTS controllers in transmission line network," in *International Journal of Advanced Engineering Research and Science*, vol. 4(1), pp. 5–11, 2017.

[18] F.M. Albatsh et al., "Enhancing power transfer capability through flexible AC transmission systems devices: A review," in *Frontiers of Information Technology and Electronic Engineering*, pp. 1–20, 2015.
[19] J.S. Masroor et al., "Comparison of UPFC and TCSC FACTS devices for system transient stability in two-area power system," in *International Journal of Applied Engineering Research*, vol. 12(1), pp. 701–705, 2017.
[20] G. Shahgholian et al., "Operation, modeling, control and applications of static synchronous compensator: A review," in *2010 Conference Proceedings IPEC*, IEEE, pp. 420–425, 2010.
[21] P.P. Raut and V. Reddy, "Improvement of real power flow control in transmission system using TCSC," in *International Journal of Advanced Research in Science and Engineering*, vol. 6(7), pp. 597–604, 2017.
[22] M. Gupta et al., "Mitigation congestion in a power system and role of FACTS devices," in *Hindawi-Advances in Electrical Engineering*, pp. 1–8, 2017.
[23] M. Gupta et al., "Mitigating congestion in a power system and role of FACTS devices," in *Advances in Electrical Engineering (Hindawi)*, pp. 1–8, 2017.
[24] R. Jena et al., "Analysis of voltage and loss profile using various FACTS devices," in *International Conference on Signal Processing, Communication, Power and Embedded System (SCOPES)*, pp. 787–792, 2016.
[25] J. Lakkireddy et al., "Steady-state voltage stability enhancement using shunt and series FACTS devices," in *Clemson University Power Systems Conference (PSC)*, pp. 1–5, 2015.
[26] S. Varma, "FACTS devices for stability enhancements," in *International Conference on Green Computing and the Internet of Things (ICGCIoT)*, pp. 69–74, 2015.
[27] A. Siddique et al., "A comprehensive study on FACTS devices to improve stability and power flow capability in power system," in *IEEE Asia Power and Energy Engineering Conference (APEEC), Chengdu*, pp. 199–205, 2019.
[28] S.R. Inkollu and V.R. Kota, "Optimal setting of FACTS devices for voltage stability improvement using PSO, GSA algorithm," in *Engineering Science and Technology (Elsevier)*, vol. 19, pp. 1166–1176, 2016.
[29] Y. Mehmet et al., "Investigation of the effects of FACTS devices on the voltage stability of power systems," in *6th IEEE International Conference on Renewable Energy Research and Applications*, IEEE, pp. 1080–1085, 2021.
[30] A.A. Ahmad and R. Sirjani, "Optimal placement of sizing of multitype FACTS devices in power systems using metaheuristic optimisation techniques: An updated review," in *Ain Shams Engineering Journal*, pp. 1–18, 2019.
[31] F.F. Moghaddam et al., "Curved space optimization for allocation of SVC in a large power system," in *Proceedings of the 6th WSEAS International Conference on Applications of Electrical Engineering, Istanbul, Turkey*, pp. 59–64, 2007.
[32] S.V. Padmavathi et al., "A. Modelling and simulation of static var compensator to enhance the power system security," in *Proceedings of the 2013 IEEE Asia Pacific Conference on Postgraduate Research in Microelectronics and Electronics (PrimeAsia)*, pp. 52–55, 2013.
[33] R. Etemad et al., "Optimal location and setting of TCSC under single-line contingency using mixed integer nonlinear programming," in *Proceedings of the 2010 IEEE 9th International Conference on Environment and Electrical Engineering (EEEIC), Prague, Czech Republic*, pp. 250–253, 2010.
[34] M. Saiveerraju et al., "DE based optimal power flow for location of UPFC considering voltage stability," in *International Journal of Recent Trends in Engineering*, vol. 2(5), 2009.
[35] A. Yousefi, "Congestion management using demand response and FACTS devices," in *International Journal of Electrical Power Energy Systems*, vol. 37(1), pp. 78–85, 2012.

[36] Y.O. Yang, "TCSC allocation based on line flow-based equations via mixed-integer programming," in *IEEE Transactions on Power Systems*, vol. 22(4), pp. 2262–2269, 2007.
[37] M. Singh and S. Gupta, "Optimal placement of FACTS devices in power system for power quality improvement," in *International Journal of Recent Technology and Engineering*, vol. 7(6), pp. 605–610, 2019.
[38] T.N. Aung et al., "Line stability index based optimal placement of UPFC," in *International Journal of Energy and Power Engineering*, vol. 6(4), pp. 47–52, 2017.
[39] K.S.L. Lavanya and P.S. Rani, "A review on optimal location and parameter settings of FACTS devices in power systems era: Model methods," in *International Journal for Modern Trends in Science and Technology*, vol. 2(11), pp. 1–8, 2016.
[40] P.S. Chindhi et al., "A comprehensive survey for optimal location and coordinated control techniques for FACTS controllers in power system environments and applications," in *IOSR Journal of Electronics and Communication Engineering*, pp. 5–11, 2016.
[41] A.S.A. Telang and P.P. Bedekar, "Application of voltage stability indices for proper placement of STATCOM under load increase scenario," in *International Journal of Energy and Power Engineering*, vol. 10(7), pp. 998–1003, 2016.
[42] G. Anju, "Optimal placement of FACTS devices using AI tools," Ph.D. dissertation, Department of Electrical Engineering, YMCA University of Science and Technology, 2013.
[43] I. Ullah et al., "Analysis of the power network for line reactance variation to improve total transmission capacity," in *Energies*, vol. 9, pp. 936–945, 2016.
[44] T. Kang et al., "A hybrid approach for power system security through the optimal installation of flexible AC transmission systems (FACTS)," in *Energies*, vol. 10(9), pp. 1–32, 2017.
[45] S. Mirjalili and A. Lewis, "The whale optimization algorithm," in *Advances in Engineering Software (Elsevier)*, pp. 51–67, 2016.
[46] P. Prajuj et al., "Power system stability enhancement using optimal placement of TCSC via genetic algorithm," in *International Journal of Science and Research*, vol. 3(1), 2017.
[47] B. Reka and R.D. Leela, "Optimal location of multiple types of FACTS devices using particle swarm optimisation," in *International Journal of Innovations in Engineering and Technology*, vol. 7(4), pp. 396–407, 2016.
[48] A.R. Jordehi, "Brainstorm optimisation algorithm (BSOA): An efficient algorithm for finding optimal location and setting of FACTS devices in electric power systems," in *International Journal of Electrical Power Energy Systems*, vol. 69, pp. 48–57, 2015.
[49] M. Ebeed et al., "Optimal setting of STATCOM based on voltage stability improvement and power loss minimisation using moth flame algorithm," in *Eighteenth International Middle East Power Systems Conference (MEPCON), Cairo*, pp. 815–820, 2016.
[50] P.K. Nguyen et al., "Cuckoo search algorithm for optimal placement and sizing of static VAR compensator in large-scale power systems," in *Journal of Artificial Intelligence and Soft Computing Research*, pp. 59–68, 2016.
[51] A. Safari et al., "Optimal setting and placement of FACTS devices using strength pareto multiobjective evolutionary algorithm," in *Journal of Central South University*, pp. 829–839, 2017.
[52] M. Musa et al., "Dynamic analysis of power loss minimization and enhancement of voltage profile using the UPFC device with differential evolution algorithm," in *European Journal of Advances in Engineering and Technology*, vol. 4(1), pp. 36–43, 2017.
[53] H.A. Devi and S. Padma, "Power system security enhancement using optimal placement and parameter setting of multi-FACTS devices with BBO algorithm," in *International Journal of Pure and Applied Mathematics*, vol. 118(5), pp. 785–804, 2018.
[54] K.S. Rajat and K.G. Vikash, *Comparison of GSA and PSO-Based Optimisation Techniques for the Optimal Placement of Series and Shunt FACTS Devices in a Power System*, Springer, pp. 375–385, 2017.

[55] A.V. Gawande and C.W. Jadhao, "Placement of FACTS devices using soft computing technique: A review," in *International Journal of Engineering Research in Computer Science and Engineering*, vol. 5(2), pp. 545–548, 2018.
[56] V.M. Chary and J. Amarnath, "Complex neural network approach to optimal location of FACTS devices for transfer capability enhancement," in *APRN Journal of Engineering and Applied Sciences*, vol. 5(1), pp. 21–25, 2010.
[57] W. Hua et al., "Stochastic environmental and economic dispatch of power systems with virtual power plant in the energy and reserve markets," in *International Journal of Smart Grid and Clean Energy*, vol. 5, pp. 25–36, 2018.
[58] J.A. Devi and K. Niramathy, "Optimal location and sizing of multi-type FACTS devices using the grey wolf optimisation technique," in *International Journal for Scientific Research and Development*, vol. 3(3), pp. 6–9, 2015.
[59] S. Ranganathan et al., "Multi-objectives based on self-adaptive firefly algorithm for multi-type FACTS placement," in *IET Journals*, vol. 10(11), pp. 2576–2584, 2016.
[60] S. Das et al., "Optimal placement of TCSC and SVC for voltage profile enhancement and minimising loss minimization using bee colony," in *IEEE Transactions on Power Systems*, pp. 978–983, 2016.
[61] S. Dutta et al., "Unified power flow controller based reactive power dispatch using oppositional krill herd algorithm," in *International Journal of Electrical Power & Energy Systems*, vol. 80, pp. 10–25, 2016.
[62] Z.A. Hamid et al., "Optimal improvement in voltage stability under contingencies using flower pollination algorithm and the thyristor controlled series capacitor," in *Indonesian Journal of Electrical Engineering and Computer Science*, vol. 12(2), pp. 497–504, 2018.
[63] V. Rao and G.V.N. Kumar, "Optimal location based on sensitivity analysis and tuning of static VAR compensator using the firefly algorithm," in *Indian Journal of Science and Technology*, vol. 7(8), pp. 1201–1210, 2014.
[64] K.M. Dosoglu et al., "Symbiotic organisms search optimization algorithm for economic/emission dispatch problem in power systems," in *Neural Computing and Applications*, vol. 29(3), pp. 721–727, 2018.
[65] P. Zakian and A. Kaveh, "Economic dispatch of power system using an adaptive charged system search algorithm," in *Applied Soft Computing Journal*, pp. 1–48, 2018.
[66] F. Khandani et al., "Optimal allocation of SVC to enhance total transfer capability using hybrid genetics algorithm," in *8th Electronic Engineering Electronics Computer Telecommunications and Information Technology Association Thailand, Conference*, pp. 861–864, 2011.
[67] S. Cui et al., "Distributed auction optimization algorithm for the nonconvex economic dispatch problem based on the gossip communication mechanism," in *International Journal of Electrical Power Energy Systems*, vol. 95, pp. 417–426, 2018.
[68] S. Chansareewittaya, "Optimal power flow for enhance TTC with optimal number of SVC by using improved hybrid TSSA," in *ECTI Transactions of Computational Information Technology*, vol. 13(1), pp. 9–20, 2019.
[69] D. Karthikaikannan and G. Ravi, "Optimal location and setting of FACTS devices for reactive power compensation using the harmony search algorithm," in *Journal for Control, Measurement, Electronics, Computing and Communications*, pp. 881–892, 2016.
[70] R.R. Verma et al., "Optimisation technique based for optimal location of FACTS devices," in *International Journal of Interdisciplinary Innovative Research and Development*, vol. 1(2), pp. 71–76, 2017.
[71] H.X. Rao and J. Huang, "A novel algorithm for economic load of power systems," in *Neuro-Computing*, vol. 171, pp. 1454–1461, 2016.
[72] A. Naganathan and V. Ranganathan, "Improving voltage stability of power system by optimal location of FACTS devices using bioinspired algorithms," in *Circuits and Systems Journal*, pp. 805–813, 2016.

[73] I. Niazazari et al., "A novel economic dispatch in power grids based on an enhanced firework algorithm," in *European Journal of Electrical and Computer Engineering*, vol. 3(4), pp. 1–5, 2019.
[74] N.S. Haroon et al., "Multiple fuel machine power economic dispatch using stud differential evolution," in *Energies Journal*, pp. 1–20, 2018.
[75] E. Purwoharjono and E.D. Marindani, "Optimal placement of UPFC using linear decreasing inertia weight-gravitational search algorithm," in *International Journal of Engineering Research and Technology*, vol. 6(1), pp. 336–342, 2017.
[76] P. Chaudhari et al., "Optimal location of FACTS devices for power system security improvement using hybrid GA-ACO," in *International Research Journal of Engineering and Technology*, vol. 4(12), pp. 551–555, 2017.
[77] A. Dihem et al., "Solving smooth and non-smooth economic dispatch using water cycle algorithm," in *5th International Conference on Electrical Engineering*, pp. 1–6, 2017.
[78] E.S. Ali and S.M. Elazim, "Mine blast algorithm for environmental economic load dispatch with valve loading effect," in *Neural Computing and Applications*, pp. 1–10, 2016.
[79] A. Ezugwu et al., "A conceptual comparison of several metaheuristic algorithms on continuous optimisation problems," in *Neural Computing and Applications (Springer)*, vol. 32, pp. 6207–6251, 2019.
[80] N.O. Keene, *The Log Transformation Is Special*, Department of Medical Statistics, Glaxo Research and Development Limited, pp. 1–25, 2015.
[81] K. Benoit, *Linear Regression Models with Logarithmic Transformations*, Methodology Institute, London School of Economics, 2016.
[82] C. Feng et al., "Log transformation and its implications for data analysis," in *Shanghai Archives of Psychiatry*, Department of Biostatistics and Computational Biology, University of Rochester, vol. 26(2), pp. 105–109, 2016.
[83] M.A. Arino and P.H. Frances, *Forecasting the Levels of Vector Autoregressive Log-Transformed Time Series*, Econometric Institute, Universidad de Navarra, pp. 10–45, 2017.
[84] Pearson Education, *Applications of the Exponential and Natural Logarithmic Functions*, Pearson Education Limited, 2018.
[85] B.M. Perry, "The exponentially weighted moving average," in *Wiley Encyclopedia of Operations Research and Management Science*, Wiley, pp. 1–9, 2017.
[86] N.I. Nwulu et al., "Microgrid energy and reserve management incorporating prosumer behind the metre resources," in *IET Renewable Power Generation*, vol. 12(8), pp. 910–919, 2018.
[87] S. Gbadamosi et al., "Multi-objective optimisation for composite generation and transmission expansion planning considering offshore wind power and feed in tariffs," in *IET Renewable Power Generation*, vol. 12(14), pp. 1687–1697, 2018.
[88] X. Hu, "Reference research of red tide based on improved FCM," in *Mathematical Problems in Engineering*, pp. 1–8, 2015.
[89] J.R. Lovvon et al., "Modelling underwater visual and filter feeding by planktivorous shearwaters unusual sea conditions," in *Ecology Society of America*, vol. 82(8), pp. 2342–2356, 2001.
[90] H.J. Jeong, "Feeding by phototrophic red-tide dinoflagellates: Five newly revealed and six species previously known species to be mixotrophic," in *Aquatic Microbial Ecology*, vol. 40, pp. 133–150, 2005.
[91] B. Walles, "The role of ecosystem engineers in the ecomorphological development of intertidal habitats," Ph.D. dissertation, Wageningen University, 2015.
[92] S. Mirjalili, *Multi-Objective Ant Lion Optimizer*, Springer, 2019.
[93] L. Daniel and K.T. Chaturvedi, "Economic load dispatch using ant lion optimisation," in *International Journal of Engineering Trends and Technology*, vol. 67(4), pp. 81–84, 2019.
[94] Z. Faisal et al., "A lion optimiser for optimum economic dispatch considering demand response as a visual power plant," in *Electric Power Components and Systems*, pp. 1–15, 2019.

[95] H.A. Devi et al., "Optimal placement of FACTS devices based on whale optimisation algorithm for power system security enhancement," in *International Journal of Innovative Technology and Exploring Engineering*, vol. 9(3), pp. 908–916, 2020.
[96] I.S. Shahbudin et al., "FACTS device installation in transmission system using whale optimisation algorithm," in *Bulletin of Electrical Engineering and Informatics*, pp. 29–38, 2019.
[97] I.S. Shahbudin et al., "Installation of the FACTS device in transmission system using whale optimisation algorithm," in *Bulletin of Electrical Engineering and Informatics*, vol. 8(1), pp. 30–38, 2019.
[98] S.S. Danish et al., "A recap of voltage stability indices in the past three decades," in *Energies Journal*, vol. 12, pp. 1–18, 2019.
[99] S. Asala and A. Gorzin, "Comparison of voltage stability indicators in distribution systems," in *Indian Journal of Science*, vol. 2(1), pp. 5–10, 2014.
[100] K.R. Vadivelu and G.V. Marutheswar, "Fast voltage stability index based optimal reactive power planning using differential evolution," in *International Journal of Electrical Engineering Research*, vol. 3(1), pp. 236–245, 2014.
[101] K.R. Vadivelu and G.V. Marutheswar, "Maximum loadability estimate for weak bus Identification using the fast voltage stability index in a power transmission system by real-time approach," in *International Journal of Electrical and Electronics Engineering and Telecommunications*, vol. 3(1), pp. 388–398, 2014.
[102] T.F. Coleman et al., *Optimisation Toolbox—for Use with MATLAB_, User Guide, Version 2*, Math Works, Inc., 1999.
[103] S. Raj and B. Bhattacharyya, "Optimal placement of TCSC and SVC for reactive power planning using whale optimisation algorithm," in *Swarm and Evolution Computation Base Data Journal*, pp. 1–30, 2016.
[104] M.S. Grewal and A.P. Andrews, *Kalman Filtering: MATLAB Theory and Practice*, John Wiley and Sons, Inc., 2001.
[105] P.J. Lagace, *Power Flow Methods for Improving Convergence*, IEEE Industrial Electronics Society, pp. 1387–1392, 2012.
[106] IEEE Standard 3002.2–2018, "IEEE recommended practice for conducting load-flow studies and analysis of industrial and commercial power systems," in *IEEE Xplore*, vol. 10, pp. 1–73, 2018.
[107] M. Mbae and N. Nwulu, "Filter feeding allogenic engineering optimization algorithm for economic dispatch," in *International Journal of Energy Production and Management*, vol. 6(2), pp. 113–128, 2021.
[108] M. Mbae and N. Nwulu, "Impact of hybrid FACTS devices on the stability of Kenyan power system," in *International Journal of Electrical and Computer Engineering*, vol. 12(1), pp. 12–21, 2021.
[109] Mathworks Inc., *The Whale Optimization Algorithm*, Mathworks Inc., 2018.
[110] S. Mirjalili, *The Whale Optimization Algorithm Source Codes*, Mathworks Inc., 2016.
[111] M.R. Al-Masud et al., "Capacity enhancement and voltage stability improvement of power transmission line by series compensation," in *EAI-Endorsed Transactions on Energy Web*, pp. 1–10, 2019.

Addendum: Material Constants and Other Parameters

A1: IEEE 39-BUS TEST SYSTEM

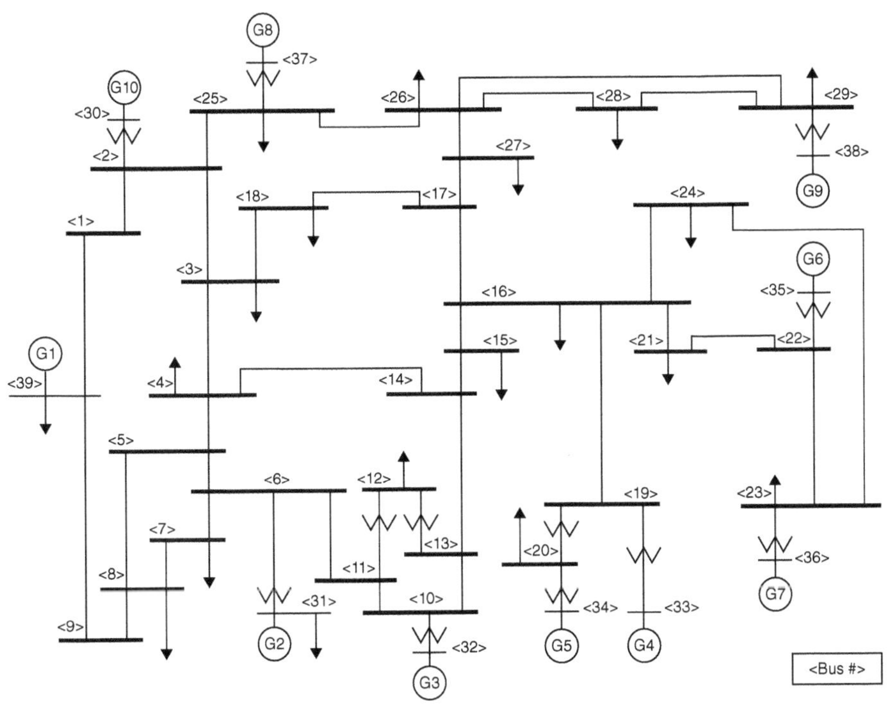

FIGURE 3.A.1 IEEE 39-Bus Test System

TABLE A.1
IEEE 39-Bus, 10-Generator, Lines, and Transformer Data

Line Data					Transformer Tap	
From Bus	To Bus	R	X	B	Magnitude	Angle
1	2	0.0035	0.0411	0.6987	0	0
1	39	0.001	0.025	0.75	0	0
2	3	0.0013	0.0151	0.2572	0	0
2	25	0.007	0.0086	0.146	0	0
3	4	0.0013	0.0213	0.2214	0	0
3	18	0.0011	0.0133	0.2138	0	0
4	5	0.0008	0.0128	0.1342	0	0
4	14	0.0008	0.0129	0.1382	0	0
5	6	0.0002	0.0026	0.0434	0	0
5	8	0.0008	0.0112	0.1476	0	0
6	7	0.0006	0.0092	0.113	0	0
6	11	0.0007	0.0082	0.1389	0	0
7	8	0.0004	0.0046	0.078	0	0
8	9	0.0023	0.0363	0.3804	0	0
9	39	0.001	0.025	1.2	0	0
10	11	0.0004	0.0043	0.0729	0	0
10	13	0.0004	0.0043	0.0729	0	0
13	14	0.0009	0.0101	0.1723	0	0
14	15	0.0018	0.0217	0.366	0	0
15	16	0.0009	0.0094	0.171	0	0
16	17	0.0007	0.0089	0.1342	0	0
16	19	0.0016	0.0195	0.304	0	0
16	21	0.0008	0.0135	0.2548	0	0
16	24	0.0003	0.0059	0.068	0	0
17	18	0.0007	0.0082	0.1319	0	0
17	27	0.0013	0.0173	0.3216	0	0
21	22	0.0008	0.014	0.2565	0	0
22	23	0.0006	0.0096	0.1846	0	0
23	24	0.0022	0.035	0.361	0	0
25	26	0.0032	0.0323	0.513	0	0
26	27	0.0014	0.0147	0.2396	0	0
26	28	0.0043	0.0474	0.7802	0	0
26	29	0.0057	0.0625	1.029	0	0
28	29	0.0014	0.0151	0.249	0	0
12	11	0.0016	0.0435	0	1.006	0
12	13	0.0016	0.0435	0	1.006	0
6	31	0	0.025	0	1.07	0
10	32	0	0.02	0	1.07	0
19	33	0.0007	0.0142	0	1.07	0

TABLE A.1 *(Continued)*
IEEE 39-Bus, 10-Generator, Lines, and Transformer Data

Line Data					Transformer Tap	
From Bus	To Bus	R	X	B	Magnitude	Angle
20	34	0.0009	0.018	0	1.009	0
22	35	0	0.0143	0	1.025	0
23	36	0.0005	0.0272	0	1	0
25	37	0.0006	0.0232	0	1.025	0
2	30	0	0.0181	0	1.025	0
29	38	0.0008	0.0156	0	1.025	0
19	20	0.0007	0.0138	0	1.06	0

TABLE A.2
IEEE 39-Bus, 10-Generator Power and Voltage Set Point Data

		Voltage	Load		Generator		
Bus	Type	[PU]	MW	MVar	MW	MVar	Unit No.
1	PQ	-	0	0	0	0	
2	PQ	-	0	0	0	0	
3	PQ	-	322	2.4	0	0	
4	PQ	-	500	184	0	0	
5	PQ	-	0	0	0	0	
6	PQ	-	0	0	0	0	
7	PQ	-	233.8	84	0	0	
8	PQ	-	522	176	0	0	
9	PQ	-	0	0	0	0	
10	PQ	-	0	0	0	0	
11	PQ	-	0	0	0	0	
12	PQ	-	7.5	88	0	0	
13	PQ	-	0	0	0	0	
14	PQ	-	0	0	0	0	
15	PQ	-	320	153	0	0	
16	PQ	-	329	32.3	0	0	
17	PQ	-	0	0	0	0	
18	PQ	-	158	30	0	0	
19	PQ	-	0	0	0	0	
20	PQ	-	628	103	0	0	
21	PQ	-	274	115	0	0	
22	PQ	-	0	0	0	0	
23	PQ	-	247.5	84.6	0	0	
24	PQ	-	308.6	−92	0	0	

(Continued)

TABLE A.2 *(Continued)*
IEEE 39-Bus, 10-Generator Power and Voltage Set Point Data

		Voltage	Load		Generator		
Bus	**Type**	**[PU]**	**MW**	**MVar**	**MW**	**MVar**	**Unit No.**
25	PQ	-	224	47.2	0	0	
26	PQ	-	139	17	0	0	
27	PQ	-	281	75.5	0	0	
28	PQ	-	206	27.6	0	0	
29	PQ	-	283.5	26.9	0	0	
30	PV	1.0475	0	0	250	-	Gen10
31	PV	0.982	9.2	4.6	-	-	Gen2
32	PV	0.9831	0	0	650	-	Gen3
33	PV	0.9972	0	0	632	-	Gen4
34	PV	1.0123	0	0	508	-	Gen5
35	PV	1.0493	0	0	650	-	Gen6
36	PV	1.0635	0	0	560	-	Gen7
37	PV	1.0278	0	0	540	-	Gen8
38	PV	1.0265	0	0	830	-	Gen9
39	PV	1.03	1104	250	1000	-	Gen1

A2: KENYA'S TRANSMISSION NETWORK DATA

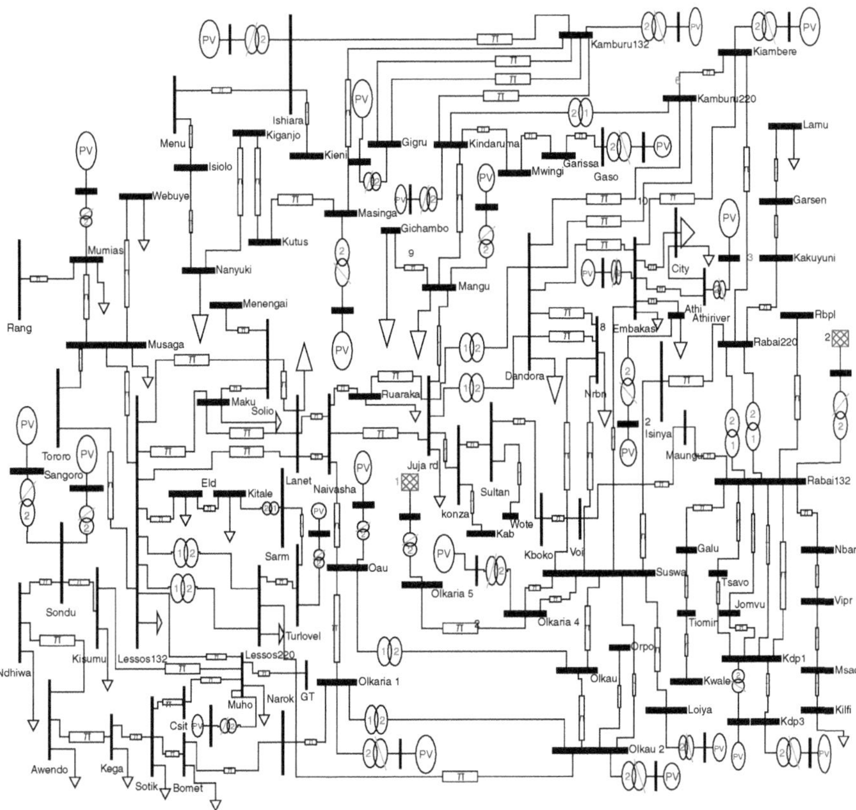

FIGURE 3.A.2 Kenya's 87–Bus, 25-Generator Data 132 KV and 220 KV Transmission Network Modeled in MATLAB's Simulink

TABLE A.3
Kenya's 132 KV and 220 KV Transmission Line Data

From	To	Line R (pu)	Line X (pu)	Susceptance (p.u.)	Rating (MVA)	Length (KM)
Kindaruma 132.00	Kamburu 132.00	0.019	0.0452	0.0086	81	18.4
Gitaru I 132.00	Kamburu 132.00	0.008	0.0186	0.0036	73	7.7
Gitaru II 132.00	Kamburu 132.00	0.0078	0.0186	0.0036	73	7.7
Kamburu I 132.00	Masinga 132.00	0.0076	0.0456	0.0084	150	18.4
Kamburu II 132.00	Meru 132.00	0.051217	0.307304	0.056609	150	124
Masinga 132.00	Kutus 132.00	0.05305	0.1071	0.0208	73	44.25
Olkaria I 132.00	Olkaria II 132.00	0.0006	0.00357	0.00508	150	4
Olkaria I 132.00	Domes 132.00	0.002519	0.014519	0.014519	150	6
Olkaria I 132.00	Naivasha 132.00	0.0097	0.0559	0.0108	150	23.1
Ulu 132.00	Juja rd 132.00	0.06455	0.1534	0.02875	81	62.5
Ulu 132.00	Sultan Hamud 132.00	0.06455	0.1534	0.02875	81	62.5
Kipevu I 132.00	Rabai 132.00	0.0204	0.0411	0.008	73	17
Kipevu II 132.00	Rabai 132.00	0.0204	0.0411	0.008	81	17
Kipevu III 132.00	Rabai 132.00	0.0105	0.0431	0.0085	150	18
Manyani 132.00	Mtito Andei 132.00	0.04645	0.11055	0.0207	81	45
Manyani 132.00	Voi 132.00	0.04645	0.11055	0.0207	81	45
Mangu 132.00	Juja rd 132.00	0.047505	0.112992	0.021606	81	46
Mangu 132.00	Kindaruma 132.00	0.109949	0.261516	0.050006	81	106.47
Juja rd 132.00	Dandora 132.00	0.0008	0.0048	0.0009	150	2
Juja rd I 132.00	Dandora 132.00	0.0008	0.0048	0.0009	150	2
Juja rd 132.00	Ruaraka Tee 132.00	0.0058	0.012	0.0023	73	5
Juja rd II 132.00	Ruaraka Tee 132.00	0.0058	0.012	0.0023	73	5
Samburu 132.00	Maungu 132.00	0.04425	0.10555	0.01975	81	43
Samburu 132.00	Mariakani 132.00	0.04425	0.10555	0.01975	81	43
Kipevu IV 132.00	Rabai 132.00	0.0105	0.0431	0.0085	150	18
Kokotoni 132.00	Rabai 132.00	0.01245	0.0255	0.000495	73	10.5
Kokotoni 132.00	Mariakani 132.00	0.01245	0.0255	0.000495	73	10.5
Rabai 132.00	Tsavo 132.00	0.0295	0.05943	0.01146	73	24.6
Rabai 132.00	Galu 132.00	0.05996	0.12079	0.02329	73	50
Eldoret 132.00	Lessos 132.00	0.0385	0.0779	0.015	73	32.1
Muhoroni 132.00	Kisumu 132.00	0.0581	0.1177	0.0225	73	48.5
Muhoroni 132.00	Chemosit 132.00	0.0368	0.0745	0.0143	73	30.7
Muhoroni 132.00	Lessos 132.00	0.068	0.1376	0.0266	73	56.7
Kisumu 132.00	Sondu 132.00	0.041667	0.119722	0.023611	150	50
Chemosit 132.00	Kisii 132.00	0.05	0.143666	0.028333	150	60
Webuye 132.00	Musaga 132.00	0.0216	0.0437	0.0084	73	18
Kiganjo 132.00	Nanyuki 132.00	0.0617	0.1246	0.0242	73	51.5
Kiganjo 132.00	Kutus 132.00	0.05305	0.1071	0.0208	73	44.25
Kilifi 132.00	Bamburi 132.00	0.0583	0.11744	0.02265	73	48.6
Owen Falls I 132.00	Tororo 132.00	0.1618	0.2755	0.0515	73	112

TABLE A.3 *(Continued)*
Kenya's 132 KV and 220 KV Transmission Line Data

From	To	Line R (pu)	Line X (pu)	Susceptance (p.u.)	Rating (MVA)	Length (KM)
Owen Falls II 132.00	Tororo 132.00	0.1618	0.2755	0.0515	73	112
Tororo I 132.00	Musaga I 132.00	0.0845	0.171	0.0329	73	70.5
Tororo II 132.00	Musaga II 132.00	0.0845	0.171	0.0329	73	70.5
Musaga I 132.00	Lessos 132.00	0.0784	0.1585	0.0308	73	66
Musaga II 132.00	Lessos 132.00	0.0784	0.1585	0.0308	73	66
Musaga 132.00	Mumias 132.00	0.0324	0.0648	0.0135	81	27
Lessos I 132.00	Kapsabet 132.00	0.035981	0.072804	0.014019	73	30
Lessos II 132.00	Kabarnet 132.00	0.027294	0.157295	0.03039	150	65
Lessos III 132.00	Makutano 132.00	0.0685	0.15215	0.0296	73	63
Lessos III 132.00	Makutano 132.00	0.0685	0.15215	0.0296	73	63
Lanet 132.00	Naivasha 132.00	0.0803	0.1626	0.0315	73	67
Lanet I 132.00	Naivasha 132.00	0.0803	0.1626	0.0315	73	67
Lanet 132.00	Soilo 132.00	0.010892	0.024192	0.004706	73	10.017
Lanet II 132.00	Soilo 132.00	0.010892	0.024192	0.004706	73	10.017
Naivasha I 132.00	Ruaraka Tee 132.00	0.0828	0.1725	0.0323	73	71.2
Naivasha 132.00	Ruaraka Tee 132.00	0.0828	0.1725	0.0323	73	71.2
Sultan Hamud132.00	Kiboko 132.00	0.0441	0.1056	0.0198	73	43
Kiboko 132.00	Mtito Andei 132.00	0.0888	0.2111	0.0395	81	86
Voi 132.00	Maungu 132.00	0.0289	0.0687	0.0129	81	28
Soilo I 132.00	Makutano 132.00	0.057609	0.127958	0.024894	73	52.983
Soilo 132.00	Makutano 132.00	0.057609	0.127958	0.024894	73	52.983
Ruaraka I Tee 132.00	Ruaraka I 132.00	0.0018	0.0036	0.0007	81	1.5
Ruaraka II Tee 132.00	Ruaraka II 132.00	0.0018	0.0036	0.0007	81	1.5
Mumias 132.00	Rangala 132.00	0.014277	0.082277	0.015896	150	34
Sondu 132.00	Sangoro 132.00	0.005993	0.012134	0.002329	73	5
Kamburu I 220.00	Kiambere 220.00	0.0074	0.0302	0.0459	250	35
Kamburu II 220.00	Gitaru 220.00	0.0014	0.008	0.0114	250	9
Kamburu III 220.00	Dandora 220.00	0.0161	0.0959	0.1364	250	107.5
Kamburu IV 220.00	Dandora 220.00	0.0164	0.0977	0.1389	250	109.5
Kiambere 220.00	Embakasi 220.00	0.0229	0.1344	0.192	250	151
Kiambere 220.00	Rabai 220.00	0.0925	0.3752	0.5767	210	440
Turkwel 220.00	Lessos 220.00	0.0458	0.1879	0.2856	210	218
Olkaria II 220.00	Nbnorth I 220.00	0.0103	0.0615	0.0875	250	69
Olkaria II 220.00	Nbnorth II 220.00	0.0103	0.0615	0.0875	250	69
Olkaria II 220.00	Olkaria III 220.00	0.001045	0.006239	0.00888	250	7
Dandora 220.00	Embakasi 220.00	0.0026	0.0108	0.0164	250	12.5
Dandora I 220.00	Embakasi 220.00	0.0026	0.0108	0.00164	250	12.5
Dandora I 220.00	Nbnorth 220.00	0.0076	0.0455	0.0647	250	51
Dandora II 220.00	Nbnorth 220.00	0.0076	0.0455	0.0647	250	51
Rabai 220.00	Malindi 220.00	0.024304	0.105738	0.16343	210	124

(Continued)

TABLE A.3 *(Continued)*
Kenya's 132 KV and 220 KV Transmission Line Data

From	To	Line R (pu)	Line X (pu)	Susceptance (p.u.)	Rating (MVA)	Length (KM)
Malindi 220.00	Garsen 220.00	0.02156	0.0938	0.14498	210	110
Garsen 220.00	Lamu 220.00	0.018522	0.080583	0.12455	210	94.5
Olkaria II 220.00	Suswa I 220.00	0.0103	0.0615	0.0875	250	64
Olkaria II 220.00	Suswa II 220.00	0.0103	0.0615	0.0875	250	65
Olkaria II 220.00	Orpower 132.00	0.0103	0.0615	0.0875	250	52
Lanet 132.00	Lessos 132.00	0.0008	0.0048	0.0009	81	125
Lanet 132.00	Lessos 132.00	0.0008	0.0048	0.0009	81	143
Juja rd 132.00	Naivasha II 132.00	0.0289	0.0687	0.0129	81	93
Juja rd 132.00	Naivasha II 132.00	0.008	0.0186	0.0036	81	93
Juja rd 132.00	Konza 132.00	0.008	0.0186	0.0036	81	86
Voi 132.00	Kiboko 132.00	0.06455	0.1534	0.02875	81	166
Maungu 132.00	Rabai 132.00	0.008	0.0186	0.0036	81	111
Rabai 132.00	Mariakani 132.00	0.06455	0.1534	0.02875	81	16
Rabai 132.00	KDP1 132.00	0.0289	0.0687	0.0129	81	68
Rabai 132.00	Jomvu 132.00	0.0828	0.1725	0.0323	73	375
Rabai 132.00	Nbam 132.00	0.0828	0.1725	0.0323	81	33
Rabai 132.00	RBPL 132.00	0.0289	0.0687	0.0129	81	26

TABLE A.4
Kenya's 132KV and 220KV Primary Sub-Station Transformers data

From	To	R (pu)	X (pu)	Rating (MVA)
Olkaria II 132.00	Olkaria II 220.00	0.003376	0.09992	90
Juja rd 132.00	Dandora 220.00	0.0002	0.105	200
Juja rd 132.00	Dandora 220.00	0.0002	0.105	200
Kiambere 132.00	Kamburu 220.00	0.002403	0.11205	270
Kiambere 132.00	Kamburu 220.00	0.002403	0.11205	270
Rabai 132.00	Rabai 220.00	0.003141	0.106893	90
Rabai 132.00	Rabai 220.00	0.003141	0.106893	180
Lessos 132.00	Lessos 220.00	0.032775	0.09999	23
Lessos 132.00	Lessos 220.00	0.032775	0.09999	23
OAU 132.00	OLKAU 220.00	0.032775	0.09999	90

TABLE A.5
Kenya's Power System Generation data

Station	Installed Capacity (MW)	Current Output (MW)	R (pu)	X (pu)	Qmax (MVAr)
Hydro					
Kindaruma	72.0	70.5	0	0.14	−10
Sondu	60.0	60	0	0.14	−4.6
Gitaru	226.0	216	0	0.13	7.3
Kamburu	94.0	63	0	0.17	33.9
Masinga	40.0	26	0	0.17	10.8
Kiambere	164.0	164	0	0.17	−8.8
Turkwel	106.0	105	0	0.18	11.5
Sangoro	21.0	20	0	0.12	45.9
Tana	20.0	20	0	0.13	7.3
Geothermal					
Olkaria I	206.0	247	0	0.18	177.2
Olkaria II	105.0	97	0	0.18	80.5
Olkaria IV	140.0	140	0	0.18	90.2
Olkaria V	316.0	305	0	0.18	195.7
Gas					
Muhoroni gas	55.0	28	0	0.11	27.2
Wind					
Ngong wind	25.5	15	0	0.16	0
Lake Turkana	310.0	200	0	0.16	0
Solar					
Garissa	50.0	40	0	0.13	15.9
Baggass					
Mumias	38.0	26	0	0.06	28.1
Diesel					
Kipevu I	75.0	54	0	0.17	71.5
Tsavo	74.1	74	0	0.16	65.9
Iberafrica	52.5	52.5	0	0.18	52
Rabai Power	90.0	90	0	0.18	19.5
Thika Power	87.0	87	0	0.18	34.4
Gulf	80.0	80	0	0.18	32.5
Triumph	83.4	83	0	0.18	44.1

4 Deep Penetration of Variable Renewables in Power Systems

4.1 OPPORTUNITIES AND CHALLENGES FOR THE INCREASED UPTAKE OF VARIABLE RENEWABLE ENERGY

Electric power generation is increasingly looking more like it was more than a hundred years ago when there were many isolated power generators. This has been caused by growing environmental concerns, increased cyber- and physical security concerns for large power plants, advances in renewable energy generation technologies, and a surge in the demand for electric power. In the mid-19th century, power generation was done using many small generation plants spread out near load centers. This quickly led to nationalized and centralized power generation plants largely managed by vertical power utilities. In the mid-1990s, liberalization and unbundling of the energy sector led to the creation of many horizontal utilities specializing in power generation, transmission, and distribution. The resurgence of renewable energy generation inevitably led to the return of distributed generation. In a way, the energy sector has gone full cycle despite technological advances and a growth in power demand [1–5].

Distributed generators owned by the consumer are increasingly taking the center stage in power generation, with surplus energy being channeled to the grid. The energy utility sector has come full circle. In the 1940s, many homes that were outside the national grid in the United States had to rely on own solar and wind generators. In the years between the 1940s and 1970s, interest in private renewable energy generation plants went down as power utilities became bigger and more reliable, with affordable energy prices. The global oil crisis of the 1970s and the increased global warming amid the rising cost of energy brought back the need to invest in renewable energy. We are now going back to the same thing, only that in the current scenario, excess power is pumped to the grid or stored in batteries and other forms of storage.

The world's first commercial wind turbine was constructed in 1891 by Poul la Cour. Interestingly, Dane was more than a hundred years ahead of his time as he used the generated power to produce hydrogen by electrolysis. Many years later, the world has returned to green hydrogen as a clean form of storage of renewable energy.

In the sitting, sizing, design, and construction of a wind or solar farm, a number of key considerations are taken into account. These include, but are not limited to, the following:

- Optimum energy generation,
- Minimization of noise pollution,

DOI: 10.1201/9781032665290-4

- Aesthetics to the surrounding environment,
- Operation and maintenance cost,
- Requirements for grid connection [6–10].

4.2 IMPACT AND MITIGATION OF INCREASED INTEGRATION OF THE INTERMITTENT ENERGY GRID

At the end of the year 2022, renewable energy accounted for 40% of the total installed generation capacity in the world. A total of 295 GW of renewable energy generation capacity was added globally in 2022, with 65% (192 GW) being solar and 25% (75 GW) being wind, both of which are intermittent in nature. The graph in Figure 4.1 shows the growth in the total installed capacity of solar and wind energy for the last 10 years [11].

The integration of increased intermittent energy into the stability of the grid presents challenges to the power system. There have been a number of recorded localized, national and regional outages occasioned by the power system operational challenges brought about by the intermittent nature of largely solar and wind energy. The seesaw nature of wind speeds and solar insolation can have a dramatic impact on the grid stability. As such, a number of measures have been developed to improve the flexibility of grid stability while ensuring optimal utilization of the power network.

A large number of innovations can be used for the integration of high-variable renewable energy into the grid while maintaining the required power system stability. These can be broadly grouped into four key categories, namely, the enabling technologies, business models, innovative market design, and system operation, as summarized in Figure 4.2.

Financial challenges remain a major bottleneck toward better integration of variable renewable energy into the existing grid. Lack of proper alignment between investment in the necessary generation and transmission assets is a major impediment to the better integration of the energy to the grid. In a number of cases, generation

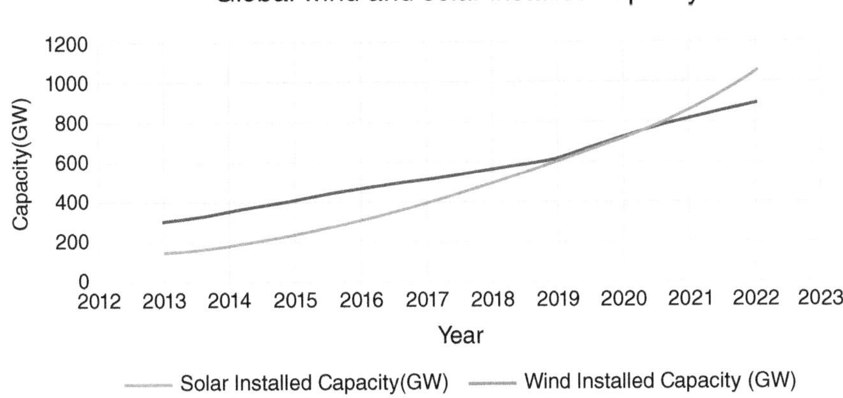

FIGURE 4.1 Global Trend in Installed Solar and Wind Energy Generation Capacity.

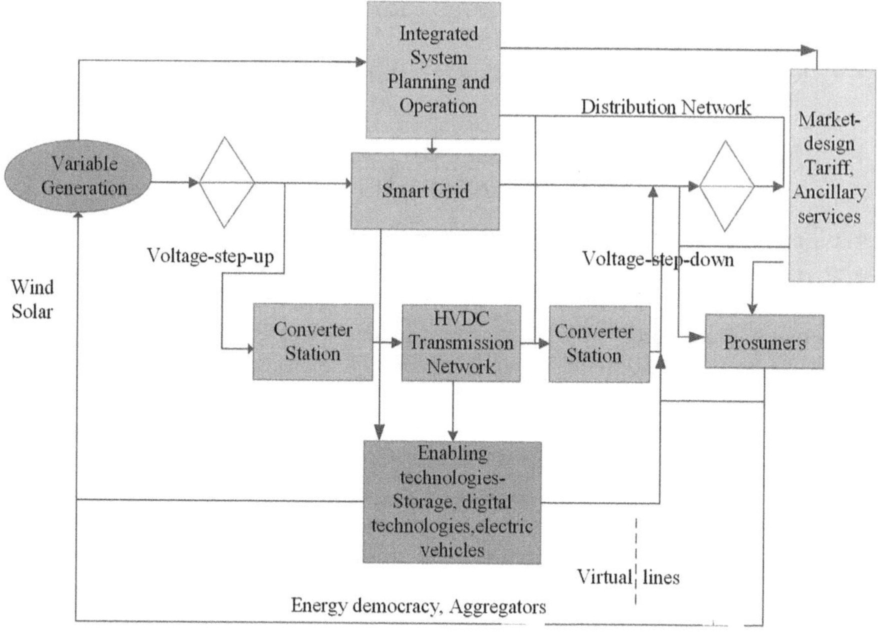

FIGURE 4.2 Mitigation Measures to Improve Power System Stability through Increased Integration of Variable Renewable Energy Resources.

plants are set up much earlier as compared to the required transmission capacity for proper power evacuation and vice versa. In both instances, the investors, grid operators, and end customers end up paying a big price. This is further complicated by the fact that most power generation assets are way off the load centers, thus the investment in transmission assets is a critical requirement. Increased transmission capacity improves grid capacity and operational flexibility, increases load balancing, reduces transmission congestion, and facilitates better utilization of additional capacity such as that of storage.

Combining of generation, transmission, and system performance planning goes a long way in ensuring that power system operators have a robust and flexible system that can respond to sudden changes in the generated energy. This is critical where power utilities have been unbundled or liberalized. Lack of synergy and industry-wide one-stop-planning shops leads to wastage of resources, increased system operational challenges, high operational cost, and non-optimization of available generation and transmission assets. Centralized static network analysis, transmission system optimization and expansion, power system stability analysis, generation adequacy, reserve capacity, and generation scheduling go a long way in developing synergy for better results in the entire value chain [12–21].

Close cooperation in the development and review of innovative rules governing how generators connect to the grid (grid codes) helps ensure system compatibility and improved power system stability and reliability. The rules should cover the entire spectrum from planning, design, and generator operation. A good example is the inclusion of the requirement for fault ride—through capability where the generators continue supplying power during faults—and the requirement for turbines to carry out reactive power support, voltage, and frequency control.

Investment in diverse generation resources spread across the country is a good way of spreading out the intermittency risk of variable renewable energy resources by giving access to a large amount of variable energy generated using a wide range of technologies spread out across various load centers. A large geographical reach also gives equal weight to weather pattern forecasting errors.

The improvement of power system operations is greatly enabled by advanced weather forecast. Granular, cloud computing-based real-time statistical, physics-based, and numerical weather forecast techniques enable power system operators to accurately determine unit commitment and spinning reserve needs. The key weather elements that require accurate forecast include wind speed, temperature, and total solar insolation. This reduces the ramping needs for expensive fossil and thermal plants and the need for expensive fast-ramping reserve capacity. Advanced weather forecasting allows for more accurate and updated forecasts every few minutes for optimal system operation and planning. Big data analytics further helps increase the accuracy of renewable energy forecast data.

The smart grid involves the deployment of a wide array of two-way communication, two-way flow of energy, and information and control technologies in transmission system automation. The use of data collection smart sensors, telemetry, the Internet of Things (IoT), the blockchain, big data analytics, virtual power plants, and smart meters will go a long way in this direction. Demand response by automated load management, including customer appliances, goes a long way in ensuring a stable grid during system disturbances. The end results are an improvement in grid efficiency, reliability, sustainability, and real-time operational flexibility and decision-making in optimal wheeling of energy.

The smart grid allows quick fault location and self-healing while preventing the cascade of major faults. More importantly, a smart power network will allow for prevention of network outages by real-time data-driven decision-making.

The bidirectional flow of energy from consumers is a great advantage offered by a smart grid. Demand-side management by, e.g., knocking off load for peak management goes a long way in improving grid stability and deferring expensive network upgrades.

The better flexibility of the smart grid allows greater penetration of highly variable renewable energy sources such as solar power and wind power, even without the addition of energy storage. The conventional power grid does not allow many distributed feed-in points from scattered generation resources. This happens while taking care of rapid fluctuations occasioned by weather changes such as clouds and varying wind speeds. This will herald the era of the much-needed smart power generation with a near-perfect match between demand and supply of energy.

Energy industry regulators need to be brought on board in order to identify and reward utilities that invest in optimized and integrated smart grid, introducing energy democracy among other benefits.

The IoT, e.g., offers a huge potential to significantly reduce unscheduled and expensive downtime by identifying problems for predictive maintenance, thereby improving reliability and reducing operational costs for generation and transmission assets. IoT-driven reliable condition-based preventive maintenance is much better than normal preventive maintenance that is scheduled based on time rather than an actual need for repairs or parts replacement. This goes a long way to reduce costly outages associated with the value chain of renewable energy assets.

Blockchain technology helps solar and wind power generators pump excess electricity into the local grid, or sell it to other customers, and receive compensation directly from the customers rather than relying on a third party. Each generating asset receives a digital identity that links to the entire production of that asset. This identity would also be linked to each owner of the corresponding credit. This record of identities and ownership would reside on the blockchain for all market participants to use. Smart contracts could then provide additional automated functionality, such as mapping energy production to carbon offset or automating credit purchasing based on a consumption profile [22–25].

Blockchain can allow distributed generation system operators to optimize grid operation by managing all connected devices through automated smart contracts, enabling flexibility and real-time pricing. Blockchain also empowers consumers to become prosumers by enabling them to monetize their excess energy by securely recording data and automatically sending and receiving payments through smart contracts.

As the technology matures, software platforms built on blockchain will be an increasingly attractive method to handle the increasingly complex and decentralized transactions between energy users, producers of various sizes, traders, utilities, and retailers. Furthermore, blockchain's ability to autonomously reconcile supply and demand between meters and computers based on smart contracts is a revolutionary efficiency improvement.

The main challenges holding back the adoption of blockchain in the energy business are legal, regulatory uncertainty, lack of trust among users, scalability of computation, privacy concerns, and cybersecurity challenges.

Regulatory issues include how to address reliability and rate-making. Regulators will have to figure out how to match supply with demand and how utilities can recover their investments. These renewable energy generators are using the infrastructure the utility has already funded to create. One way utilities could recoup their investment is through microtransactions. There could be a fee for the use of the network, and blockchain would enable utilities to charge and collect that fee. Smart contracts can help with this.

The energy industry is highly controlled and regulated in many countries. Rigid legal and regulatory barriers are a major impediment to encouraging the right investment. Energy trade across diverse utilities and countries in a regional power pool, e.g., is still a big challenge in many regions across the globe. Collaborative, innovative, and progressive laws and regulations in the bulk and retail electricity markets that encourage flexibility from market players will go a long way in this direction.

Investing in the right enabling technologies will help improve power system stability on increased integration of renewable energy. A deeper penetration of variable renewable energy needs better system flexibility to take care of intermittency and uncertain power generation. These include, but are not limited to, more flexible generating units with more granular scheduling and dispatch intervals, battery storage, pumped hydro, solar boilers, compressed air, super capacitors, superconducting magnets, hydrogen, and flywheels, among other forms of storage and demand response.

Increased investment in battery storage has been encouraged by the rapidly falling costs of batteries, the increased uptake of intermittent renewable energy, and the increased importance of storage batteries as a way of retiring unclean fossil fuel-fired generators. Energy storage systems are critical to modernizing power grids and will become a practical alternative to building new generation or network reinforcements. Storage enhances network predictability, reliability, resilience, and stability by creating buffer reserves of electricity for frequency and voltage support, system peak, increasing system reserve capacity, renewable capacity firming, and load management

Modeling of storage investment involves proper understanding of the storage technology, applicable business models, and other determinants such as infrastructure investment deferral, peak system load management, frequency regulation, energy price arbitrage, customer demand-charge management, backup power, and cost–benefit analysis.

To help utilities and storage developers navigate the new obligations, utilities and governments need to create the necessary laws, policies, incentives, energy storage roadmaps, and targets. Good examples of a progressive policy are the definition of battery storage as a generation asset and the payment for storage for ancillary services.

The use of flexible loads assists in the improvement of system flexibility by being switched off or by using low-energy pricing during peak intermittent generation to encourage utilization of excess energy. Loads such as water desalination plants, ice production plants, electric vehicles, and space heating are very good for this application. Innovations such as the granular real-time pricing model and remote load management through smart grids are critical in this regard.

The use of the zonal or nodal segregated pricing model helps manage transmission network congestion while encouraging investors to develop network resources in a balanced manner across all parts of a country or state. Negative pricing in certain jurisdictions encourages investors to move elsewhere when sitting new generation projects, thus taking care of new loads, balancing system operation, and reducing on network congestion. The differentiated tariff model encourages planning and investment in intermittent renewables that match more closely the requirements of the grid, and driving customer behavior change.

Imbalanced payment rules, whereby generators are penalized for deviating from their generation schedules, encourage the generation system owners to invest in advanced weather forecast systems and in storage.

Smart charging of electric vehicles adapts the charging process to the conditions of the real-time power system. This helps avoid the curtailment of renewable energy and, at the same time, reduce the peak demand needs. Vehicle-to-grid technologies enable power supply back to the grid on a need basis.

Regional electricity markets and power groups with harmonized rules and regulations in the wholesale market and ancillary services go a long way in better sharing of diverse renewable energy resources. Creating a regional market through the use of transmission interconnections consolidates balancing areas and increases the flexibility of the grid. When renewable energy resources are available over large regions, the need for operating reserves, as well as the curtailment requirements and costs, is reduced. The main goal of every power pool should be to harmonize bulk power markets with the aim of creating a large shared market for improved and efficient energy use across national borders. Common rules across member countries to promote efficient use of cross-border capacity and harmonization of the wholesale power market should be a key goal. Features such as day-ahead market coupling and continuous intraday trading allow cross-border trading of electricity closer to real time.

Construction of DC super-grid interconnectors with enormous capacity to wheel electric energy over very long distances in a more efficient way than AC systems should be a top priority of regional power pools. This helps contain AC power system disturbances within a given country or power utility, rather than having the same to cascade across regions.

As a result of their fast control speed, power transmission via a DC link has better control on transient and dynamic line stability, in addition to the ability to limit fault current in the DC lines.

The power transfer limit of an AC line is a function of the angle difference between the voltage phasors at the sending and receiving ends, which depends on the distance. As such, the power transfer capability of an AC line reduces with distance. This also brings in the need to invest in line compensation devices such as FACTS devices for very long and heavily loaded transmission lines.

Interconnection of two power systems through AC tie-lines requires automatic generation controllers of both systems to be well coordinated using tie-line power and frequency control signals. This is a big challenge that results in frequent line tripping, increased fault levels, and the cascade of disturbances from one network to the other.

Innovative ancillary services will go a long way toward improving the smooth integration of variable renewable energy into the grid. This is critical so as to improve system flexibility, ensure proper remuneration of each service to the grid, and encourage effective fast-ramping grid service. Opening up to new players such as battery storage, load management, and other distributed services will go a long way in encouraging the right investments [26–31].

4.3 FUTURE PERSPECTIVES ON INCREASED PENETRATION OF RENEWABLES

The renewable energy space will evolve around progressive policies and laws, innovative incentives, operating procedures, evolving business models in line with technical needs, and full evaluation of a power system's stability and reliability.

The overall share of the renewable energy injected into power grids is expected to double to around 80% of the total energy mix. The variable solar and wind power are expected to take the lion's share. As such, continuous improvement on grid stability, more deployment of Industry 4.0 technologies, smart grids, and the emergence of Society 5.0 will play a pivotal role in proper harnessing of the renewable energy resources [32–39].

4.4 SUMMARY

The growth of intermittent renewable energy generation will continue to grow across the world in the foreseeable future. The challenges with global oil prices, the need to reduce carbon emissions and global warming, and the whole sustainability challenge will continue to present opportunities in this area.

As such, there is a need to continue with more innovations, investments, and research into better ways of integration of the variable resources into the existing grids while also modernizing the said grid.

REFERENCES

[1] UN Environment, *Potential for Africa and Opportunities for Investment*, UN Environment, 2019.
[2] CIGRE, *Is Inserting Renewables into the Electric Grid Stable?* CIGRE, 2019.
[3] Innovations Accelerator, *Squaring the Circle-Grid Stability and Renewable Energy*, Innovations Accelerator, 2020.
[4] S.Mamatha and T. Tilak, "Issues challenges, causes, impacts and utilisation of renewable energy sources-grid integration," in *International Journal of Engineering Research and Applications*, vol. 4(3), pp. 636–643, 2014.
[5] S.Yasmeena and T. Das, "A review of technical issues for grid connected renewable energy sources," in *International Journal of Energy and Power Engineering*, vol. 4(5), pp. 22–32, 2015.
[6] Alexandra Von Meier, *Challenges to the Integration of Renewable Resources at High System Penetration*, Department of Energy of the USA, 2016.
[7] Greentech Media, *What Are the Impacts of High Wind and Solar Penetration on the Grid*, Greentech Media, 2021.
[8] Meteolcd, *Is There an Upper Limit for Wind-Solar Grid Penetration*, Meteolcd, 2019.
[9] Maxwell, *Stabilize the Grid Higher on Higher Penetration of Renewables*, Maxwell, 2022.
[10] International Renewable Energy Agency (IRENA), *Innovation Landscape for a Renewable-Powered Future: Solutions to Integrate Variable Renewables*, IRENA, 2019.
[11] K. Benjamin, *High Penetration of Renewable Energy: Possible Scenarios, Implications, and Best Practices from International Experience*, National Renewable Energy Laboratory, 2013.
[12] J. Cochran et al., *Global Status Report Integrating Variable Renewable Energy in Electric Power Markets: Best Practices from International Experience*, Ren21, 2019.
[13] America Physical Society, *Integrating Renewable Electricity in the Grid*, America Physical Society, 2015.
[14] Department of Energy and environment, *Challenges of Integrating Solar and Wind into the Electricity Grid*, Chalmers University of Technology, 2014.
[15] EPRI, *Demystifying the Value of Energy Storage*, EPRI, 2019.
[16] Department of Energy of the United States of America, "Energy information administration (EIA)," in *Trends in the US Battery Storage Market Trends*, Department of Energy, 2018.
[17] IRENA, *Tools to Estimate Storage Costs*, IRENA, 2021.

[18] AT Systems, *Power from Big Data: Are Europe's Utilities Ready for the Age of Data*, The Economist Intelligence Unit, 2017.

[19] O. Amonoo, *Lessons and Challenges in the Reform of the National Electricity Sector: The Case of Ghana*, The Energy Commission, 2016.

[20] L. Bird and M. Milligan, *WREF 2012: Lessons from Large-Scale Renewable Energy Integration Studies*, World Renewable Energy Forum, 2012.

[21] BMU, *Energy Concept for an Environmentally Sound, Reliable, and Affordable Energy Supply*, BMU, 2012.

[22] D. Burke and M. O'Malley, "Maximising firm wind connection to security constrained transmission networks," in *IEEE Transactions on Power Systems*, vol. 25(2), pp. 749–759, 2014.

[23] J. Cochran et al., *Integrating Variable Renewable Energy in Electric Power Markets: Best Practices from International Experience: Summary for Policy Makers*, National Renewable Energy Laboratory, 2014.

[24] Danish Electricity Infrastructure Committee, *Technical Report on the Future Expansion and Undergrounding of the Electricity Transmission Grid*, Danish Electricity Infrastructure Committee, 2014.

[25] G. Masters, *Renewable and Efficient Electric Power Systems*, Wiley-Interscience, 2004.

[26] L. Philipson and H. Lee, *Understanding Electric Utilities and Deregulation*, 2nd ed., Taylor and Francis, 2006.

[27] International Renewable Energy Agency (IRENA), *Renewable Capacity Statistics*, IRENA, 2023.

[28] J. Douglas, *High Penetration of Renewable Energy: Possible Scenarios, Implications, and Best Practices from International Experience*, National Renewable Energy Laboratory (NREL), 2013.

[29] M.M. Hand et al., *Renewable Electricity Futures Study: Exploration of High Penetration Renewable Electricity Futures*, National Renewable Energy Laboratory (NREL), 2012.

[30] L. Bird, *Integrating Variable Renewable Energy in Electricity Power Markets: Best Practices from International Experience*, National Renewable Energy Laboratory (NREL), 2012.

[31] B. Georgiana et al., *Analysing the Impacts of Renewable Energy Sources on Power System National Network Code of the Power System*, MDPI, 2017.

[32] A. Meier, *Challenges to the Integration of Renewable Resources at High System Penetration*, California Energy Commission, 2011.

[33] International Renewable Energy Agency (IRENA), *Innovation Landscape for a Renewable-Powered Future: Solutions to Integrate Variable Renewables*, IRENA, 2019.

[34] IEA-ETSAP and International Renewable Energy Agency (IRENA), *Integration of Renewable Energy in Power Grids*, IRENA, 2015.

[35] International Renewable Energy Agency (IRENA), *Smart Grids and Renewables; A Guide for Effective Implementation*, IRENA, 2013.

[36] International Energy Agency (IEA), *Harnessing Variable Renewables; A Guide to the Balancing Challenge*, IEA, 2011.

[37] L. Jones, *Strategies and Decision Support System for Integrating Variable Energy Resources in Control Centres for Reliable Grid Operations; Global Best Practices, Examples of Excellence and Lessons Learned*, Alstom Grid Inc., 2011.

[38] International Renewable Energy Agency (IRENA), *Blockchain: A New Tool to Accelerate the Global Energy Transformation*, IRENA, 2018.

[39] V. Sood, *HVDC and FACTS Controllers; Application of Static Converters in Power Systems*, Kluwer Academic Publishers, 2014.

5 The Quarters of Industrial Revolution and Power Systems

5.1 INTRODUCTION TO INDUSTRY 4.0

The Industrial Revolution has come a long way, right from Industry 1.0 that was run by water and steam. This was closely followed by the Second Industrial Revolution, the hallmark of which was the metamorphosis of knowledge and which was characterized by electric power-driven mass production of goods. The Third Industrial Revolution relied heavily on computer and other emerging digital technological advances. This, in turn, has paved way for the Fourth Industrial Revolution (4IR). The 4IR or Industry 4.0 is a revolution driven by the confluence of technologies with blurred lines between the physical, digital, and biological space, as seen in Figure 5.1 and Table 5.1.

The 4IR is impacting and disrupting every known convectional industry, profession, and business. The transformation is underpinned on the research, development, and convergence of new and emerging cutting-edge technological advances in a number of fields such as artificial intelligence (AI), cobotics, Internet of things (IoT), autonomous vehicles, 3D printing, nanotechnology, genetic engineering, augmented reality, genome editing, biotechnology, materials science, energy storage, additive manufacturing, material engineering, quantum computing, and other technologies [1–12].

The emergence of Society 5.0 is very important for maximum realization of the benefits of the 4IR in power systems. Society 5.0 is a society that balances socioeconomic human advancement with high integration of digital, physical, and biological

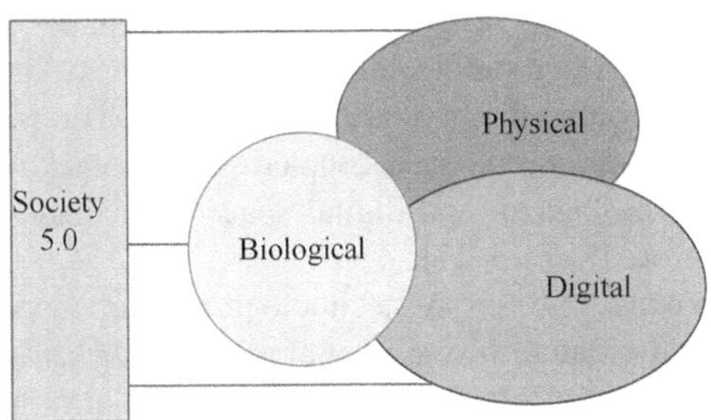

FIGURE 5.1 Illustration of the 4IR.

DOI: 10.1201/9781032665290-5

TABLE 5.1
4IR Technological Facets

Technological Facet	Field
Digital	Artificial intelligence Machine learning Internet of things Big data Cloud computing Quantum computing
Physical	3D printing Robotics Autonomous cars Advanced materials Energy storage Drones Nanotechnology
Biological	Biotechnology Biomedical engineering Genome editing Neurotechnology Bioinformatics Genetic engineering

systems in everyday life. This will be the ultimate successor of Society 1.0 (hunters and gatherers), Society 2.0 (agrarian), Society 3.0 (industrial), and Society 4.0 (information and communication).

5.2 TECHNOLOGICAL ADVANCES IN INDUSTRY 4.0

The 4IR has brought about the fastest technological advances in recorded history. There is no clear line between one technology and the next, and there is no clear-cut end of one innovation and the beginning of the next. The overlaps between one technological advancement and the next coupled with the lack of a clear distinction between the physical, biological, and digital space have made the advances more thrilling than ever before [5–7, 13–17].

Augmented and virtual realities are technologically enhanced representations of the real physical world created through the application of sensory stimulation and digital visualization. The top priorities of augmented reality are to bring out the characteristics of the given physical world, increase knowledge of those features, and obtain sensible and approachable intelligence that can be employed in real applications. This is done by making use of realistic-looking sights and objects. These technologies have brought on board the era of immersive digital experience like never witnessed before by converging the digital and physical space through a set of sensors, algorithm, cameras, goggles, and other devices. Using both virtual and augmented realities, it is possible to see virtual objects while building a feeling in which the physical and the digital spaces are one and the same.

The metaverse is a combination of the prefix meta (inferring transcendence) and the word universe, which depicts a hypothetical synthetic environment linked to the physical world. The metaverse is characterized as a virtual environment that incorporates physical and digital aspects, which has been made accessible by the combination of Internet and web technologies, as well as extended reality.

Additive manufacturing promises better, innovative, and more complex system designs, coupled with easier and more accurate fabrication processes. This relies heavily on computer-aided design, 3D printing, fast system prototyping, and less tooling, thus reducing production cost. Coupling this with nanotechnology has ushered in miniature industrial products that are cheaper and lighter and use smaller quantities of materials, reduce waste, and have a much lower energy consumption and a far much smaller carbon footprint. The invention of nano-enabled materials that zero in on the control, modification, and manipulation of materials at the atomic level will be the corner stone in industrial production processes geared toward improved manufacturing efficiency and output.

The rapidly evolving era of AI and quantum computing has brought on board extremely fast processing speeds capable of handling tons of complex data and converting them into useful patterns for better decision-making. Continual collection and processing of huge quantities of data from engineering systems collected using billions of sensors via communication systems is getting faster and better.

Computers and AI systems that mimic human intelligence are the inspiration behind robotics and cobotics. AI has given rise to other improvements such as machine learning, deep learning, soft computing, and other forms of advanced computational intelligence.

The IoT has the capability to monitor, collect, exchange, analyze, and deliver useful insights for smarter, faster, and more accurate business, technical, operational, and managerial decisions. This is through real-time data collection, analysis, monitoring, and decision-making at the click of a button. By deploying billions of sensors, appropriate communication networks, big data analytics, and computing techniques, IoT enables measurement and optimization for improved and transformative system and operational efficiency for improved productivity and minimization of costly machine downtime.

A digital twin is a digital reproduction of a real component created by combining simulations and service metadata. Data from numerous sources are included in the digital representation throughout the product life cycle. These data are constantly restructured and shown in a variety of ways in order to forecast current and future situations in both operational and design settings, and therefore improve decision-making.

The rollout of 5G technologies will play a critical role in the faster realization of the IoT capabilities. Cloud, fog, and edge computing will be major enablers to fuel other broader uses including in the IoT space. These have allowed practitioners and businesses to take full advantage of a variety of computing and data processing techniques. Edge computing allows data processing to be done locally at multiple decision points, thereby minimizing network traffic and congestion. Fog and edge computing will play a big role in helping organizations, institutions, and businesses free themselves from the need of keeping own huge infrastructure.

5.3 IMPACT OF INDUSTRY 4.0 ON POWER GENERATION, TRANSMISSION, AND DISTRIBUTION SYSTEMS

The century-old conventional one-way power grid infrastructure is heavily disrupted by the three Ds of digitization, decarbonization, and decentralization, with the aim of improving operational efficiency, business sustainability, and customer experience. The rise of decentralization has been fueled by the rapid adoption of distributed renewable energy generation technologies, both stand-alone and grid-connected generation facilities. This has also been aided by the rapid adoption of energy storage systems and micro- and mini-grids, among other forms of distributed power generation and distribution. The growing concern of decarbonization is informed by the need to reduce the power sector carbon footprint. This relies heavily on renewable energy projects, storage, electric vehicles, and virtual power plants and transmission lines. On the flip side, digitization promotes increased penetration of digital technologies such as blockchain and other AI, digital, and computing techniques [18–26].

Industrial robotics, drones, and other unmanned aerial vehicles carry out sensor-based transmission and other power network inspection functions, thereby improving diagnostic capabilities, reliability, maintenance activities, and asset optimal utilization. This is done at a much reduced cost while giving troves of useful data that are continuously analyzed. This enables the planning of proactive maintenance activities that is informed by a continuous advanced analysis of the collected data. Predictive maintenance using big data and machine learning algorithms has been used to identify potential equipment failures before the occurrence of the same.

Robotics is helping improve transformer maintenance and routine inspections. A submersible robot has been developed to inspect major oil-cooled transformers. Using cameras, it captures and wirelessly transmits internal images for continuous and real-time analysis.

In power distribution, continuous monitoring of the health of the transformer is necessary. This includes indicators such as transformer oil level, current and voltage levels, load balancing, temperature, and insulation levels, among others. This is done by the deployment of a system of sensors that are mounted on the IoT platform.

The use of IoT by power utilities to develop a reliable, condition-based proactive preventive maintenance as opposed to reactive and risky maintenance activities will go a long way in driving down the operational costs in terms of equipment repair, maintenance, and costly downtimes. The IoT proactive maintenance investment should first focus on the high-value capital assets in generation, transmission, and distribution for better and faster return on investment. The assets are continuously monitored using millions of sensors, and the collected data are used to analyze the health and impending failure of the assets, as well as optimal utilization of the same.

In asset management, IoT heralds a transition from scheduled maintenance programs to the following:

- Condition-based maintenance in which patterns from collected data are identified to plan asset maintenance;
- Predictive maintenance where trends in sensor-collected data are used to predict impending equipment failure and thus better maintenance plans;
- Risk-based maintenance in which the decision about the maintenance of an asset is also based on optimizing the use of maintenance of resources across all assets. In this case, the risk of failure is used as an indicator to allocate maintenance resources. The risk here is the product of the probability of failure and the economic impact of the said failure.

These are in contrast to reactive maintenance (after the occurrence of failure) and preventive maintenance (based on data on equipment and time usage).

This is illustrated in Figure 5.2.

The advantages of IoT-based maintenance include improved asset utilization, reliability and availability, reduced maintenance cost, reduction of unplanned costly downtime and outages, cheaper inventory of spare parts, and improved overall productivity. Other benefits include, but are not limited to, resource sharing, timely asset retirement, proper asset monitoring, optimal operation and maintenance management, procurement monitoring, and proper inventory management [27–31].

A smart grid makes use of billions of sensors and digital information and communications technologies to gather and act on information in an automated manner so as to improve power system efficiency, reliability, economics, and sustainability of the production and distribution of electric energy. The information technologies used

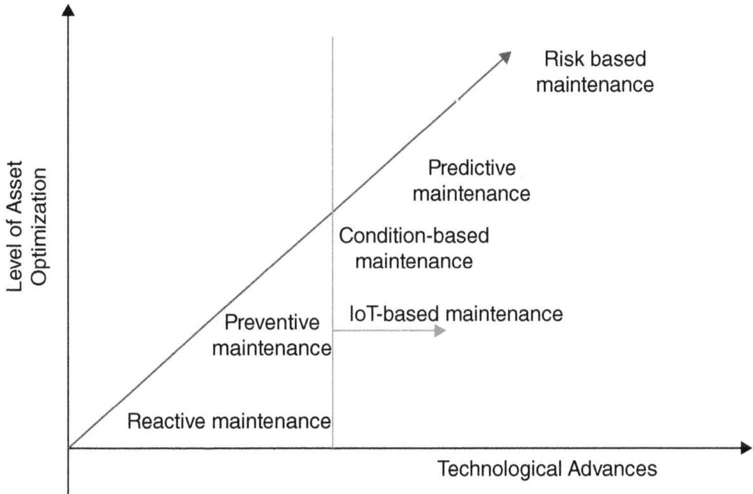

FIGURE 5.2 Classification of Asset Maintenance.

allow for visualization and conversion of large amounts of data into easily understood visual graphics and patterns for ease of decision-making.

The rise and development of the smart grid has been motivated by the need to change the conventional centralized power grid to a more robust grid that allows bidirectional flow of energy, as well as the necessary smart demand side management controls. Bidirectional flow of energy allows for distributed generation from sources such as wind and solar power. This in addition provides a robust way of handling their intermittent nature without causing grid instability. Smart grids allow fast fault location and self-healing capability while preventing the occurrence of system-wide network outages by real-time data analytics and decision-making.

Digital technologies are also being used to improve power distribution. Smart grids are being used to balance the supply and demand of power. This approach helps reduce the cost of electricity and improve the reliability of the power grid. They are used to optimize the use of power during peak demand periods. Machine learning algorithms are used to predict when power demand will be high and when it will be low. This information is then used to adjust the distribution of power to meet the needs of customers.

In a nutshell, the benefits of smart grids include, but are not limited to, the following:

- Quicker electrical restoration during power supply disturbance and outages;
- Reduction in operating, maintenance, and network running costs for power utilities;
- Shaving of peak demand through demand response, which leads to delayed expensive generation and network expansion projects;
- Better integration of intermittent renewable energy;
- More efficient and better integration of renewable energy generation from consumers.

Intermittent renewables are in a pole position to assist in addressing the slippery goals under the energy trilemma—the transition to net-zero emissions, energy security for the world, and sustainable energy pricing. The expected growth of renewable energy generation will largely be driven by emerging technologies, as illustrated in Figure 5.3.

Virtual power plants, virtual transmission and distribution lines to address grid bottlenecks, energy storage, and smart IoT-based micro-grids will make it easier to generate and distribute energy.

Smart power generation helps utility operators match electric energy production and demand by using multiple generators that can be ramped on and off while taking care of the base and peak load demands. This is a greater challenge with the continued injection of intermittent renewable energy into the grid. This is because intermittent renewables like solar and wind energy have the downside of negatively impacting on power system stability, reliability, and control.

Digital technologies have been used to improve the efficiency and reliability of power generation. This helps power companies and utilities reduce costly downtime and improve the reliability of their power generation plants. Data are continuously

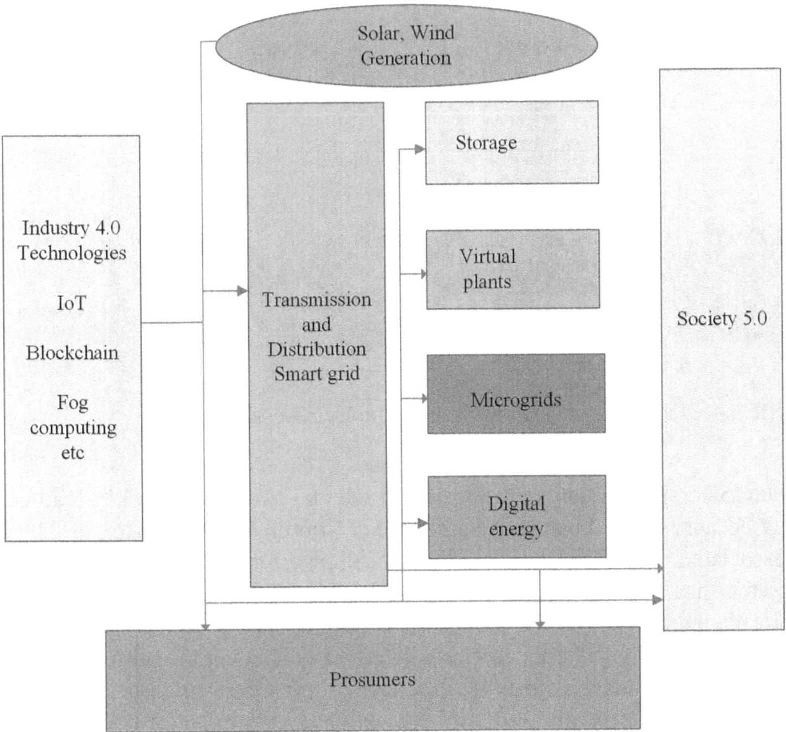

FIGURE 5.3 Industry 4.0 and the Energy Trilemma.

collected using millions of sensors installed in the right places in the machines and equipment. In other instances, drones are used to inspect equipment such as solar panels [32–38].

Real-time generation and demand monitoring via smart sensors and aggregated telemetry will aid in generation demand matching for maximum production effi ciency, which also assists in minimizing the cost of power. This is possible through Industry 4.0 technologies such as cloud and fog computing, as illustrated in Figure 5.4.

IoT generation-level technologies are designed to integrate a variety of sources of energy such as hydro, coal, gas, oil, nuclear, wind energy, and solar energy. System design presents the importance of the integration of fog and edge computing. Several sensor motes—sensors and wireless communication—will be used for continuous monitoring of the various sources of power generation.

Modern solar power generation systems have more advanced technologies to efficiently extract solar energy, including a maximum power point tracker (MPPT) system. During cloudy days, the MPPT inclines the panels directly toward the brightest part of the solar radiation of the sky. The existence of a storage plant is crucial because solar power must store the energy for distribution during peak times. The IoT assists in communicating all data from the solar system in real-time, maximum solar radiation tracking, remote supervision during preventive maintenance, and fault detection of the system.

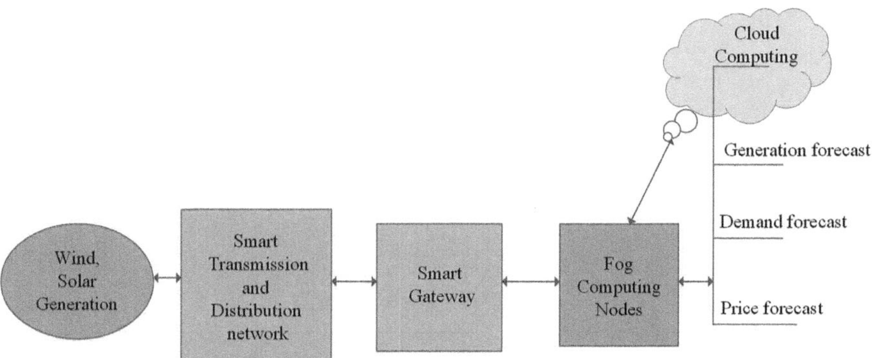

FIGURE 5.4 Industry 4.0 and Power Generation-Demand Matching.

Solar generation systems are located in remote areas far from key load centers largely because of the huge land requirement. Continuous monitoring of the panels and associated control systems is a huge challenge for system operators. As such, the system has to be equipped with a continuous real-time monitoring system to monitor the parameters of the photovoltaic system and log the necessary data on a cloud server. The logged data can be utilized for enhancing the performance of the photovoltaic system and identifying the causes of poor performance.

Digital technologies are used for more seamless integration of renewable energy sources such as solar and wind energy into the power grid. Although the unpredictable and intermittent nature of these variable energy sources can present challenges for maintaining the stability and reliability of the grid, digital technologies can support the integration process by improving the visibility of the grid and providing tools for coordination and learning. Additionally, digital technologies are being used to optimize the use of renewable energy sources. Machine learning algorithms are being used to predict when renewable energy sources will be available and when they will need to be stored for later use.

It is paramount that emerging techniques of the 4IR techniques are able to seamlessly work with existing transmission and distribution network management technologies. These include, but are not limited to, supervisory control and data acquisition (SCADA), advanced distribution management system (ADMS), energy management system (EMS), wide area measurement system (WAMS), and phase measurement unit (PMU). This is for power system operation and control, addressing grid bottlenecks, fast restoration on faults, and better integration of intermittent renewable energy [39–50].

SCADA offers real-time telemetered data on the performance of each power system node, bus bar, and equipment of the transmission line. The system allows remote monitoring, control, and operation of the various equipment such as circuit breakers, among others. Hand in hand with this, the WAMS and PMU system consist of an array of sensors, and applicable communication technologies are used to collect data on the status of various transmission system nodes and bus bars. These should fit very well with the IoT system that is designed to handle data measurement and processing.

The digital technologies-based control system is more expensive than the conventional SCADA system. However, it is more diagnostic due to the higher frequency of information and the maximum sampling rate.

The EMS utilizes the SCADA system data for real-time and continuous data analysis and decision-making. It enables better visualization, operation, optimization, and maintenance of power systems.

The ADMS platform is critical for the management and optimization of the power system distribution network. It has functions that include, but are not limited to, automated outage restoration, load transfers, fault location and sectionalization, peak demand management, and voltage control.

Digital twins generate a digital backbone across a given value chain to represent a clear path for all involved functions, thus producing a more complete end-to-end solution and understanding of the system at hand. Digital twin and other digital simulation technologies will assist in speeding up the research and innovation required for power generation equipment by creating a strong curve in the generation of wind and solar energy generation. Predictive digital twins of wind turbines and solar panels allow for the development of lighter, leaner, more efficient, and progressive designs. The digital feedback loop provided by this technology allows for improvements to both the design and operations of the generation systems, which in turn lead to lower costs, shorter timeframes to design and build generation capacity, and thus increased renewable energy adoption. The technology has greatly improved transformer and other grid asset management and operations by providing key insights, thus paving the way for real-time monitoring and proactive maintenance. By harnessing AI and machine learning, extensive data can be analyzed to anticipate potential system disruptions, ensure consistent energy output, and extend the longevity of the equipment.

Industry 4.0 tools such as augmented reality models, machine learning, and IoT have allowed power plant managers to carry out the following:

- Develop productive and resilient operations that drive fast decision-making by operating, engineering, and management teams,
- Automate and optimize systems through machine learning and other technologies to eliminate redundant duties, thus allowing employees to work safely and efficiently while focusing on higher-value goals for better productivity and innovation.

Industry 4.0 has made a major contribution to deepening the much needed energy democracy by way of bettering energy transparency, digital energy, and energy trading and providing above-board power system flexibility. This has been heralded by the energy Internet powered by emerging progressive technologies such as blockchain that allows peer-to-peer energy trade.

Digital technologies are used to improve the customer experience. Mobile apps are used to provide customers with real-time information about their energy consumption. This information helps customers better manage their energy usage, thereby reducing their power bills. In addition, digital technologies are used to improve the billing process. Digital billing and payment systems are used to reduce the cost and time associated with the billing process. Another impact of digital transformation on customers is the emergence of new business models, a good example being consumers who generate own renewable energy and pump the excess to the grid. This is possible with the emergence of a smart grid that allows a two-way flow of energy.

The smart home solution allows for the daily measurement and invoicing of energy use, as well as the visualization and display of the energy consumption of particular household appliances. This technology creates transparency and provides an opportunity to identify energy-saving potential. As most interactions can be completed through online consumer portals, these solutions increase customer satisfaction while reducing costs.

Customer satisfaction is another critical driver for going digital. Customer needs and expectations have changed over the years, and companies compete to provide better services and achieve higher satisfaction. The importance of climate-friendly energy, its use, and cost transparency has increased. Smart meters and smart homes are digital applications that can help meet the goals of decreased costs, increased transparency, and increased use of renewable energy.

In energy consumption, the main use of IoT is to optimize power usage at home by continuous monitoring of all appliances and equipment such as refrigerators, washing machines, lights, and others. An IoT-based smart grid in this field provides an opportunity to realize the concept of smart home automation. This enables an evaluation of energy efficiency and response to demand. Monitoring parameters such as temperature, humidity, and air quality enable proper decision-making. Consumers can monitor and manage their energy usage remotely at the touch of a button, e.g., remotely switching off an idle appliance in the house. Smart metering will be useful in this regard.

The integration of IoT into the advanced metering infrastructure transforms the metering infrastructure into a real-time system to obtain data on customer load. In addition to automatically responding to demanded household appliances or devices, end users may spontaneously perform certain demand response initiatives. Therefore, the user can control various devices and devices employing computer-specific interfaces or smartphones and tablets more conveniently using an IoT-based communication infrastructure.

Vehicle-to-grid (V2G) technologies are programmed and operated with the help of IoT gateways for absorbing or powering the grid. The storage unit will supply the grid at its peak, depending on the load pattern and capacity. V2Gs can be automatically charged for real-time electricity price tracking at midnight and are able to sell excess stored energy at peak times to the main grid. Thus, a smart cloud computing environment can be materialized with IoT technologies and information and communication technologies with high bandwidth and high-speed communication infrastructure.

Energy storage technologies contribute to dispatch ability by resolving the imbalances of intermittent renewable resources. A grid-scale battery energy storage system (BESS) consists of a control system, a battery bank, protective circuitry, power electronics interface for AC–DC power conversion, and a transformer to alter the BESS output to the transmission or distribution system voltage level. The IoT enables the collection of the huge amount of data required for continuous monitoring and decision-making. Energy storage systems perform roles that include, but are not limited to, frequency regulation, bulk energy time shifting, frequency stability improvement, demand response, and power flexibility and reliability improvement.

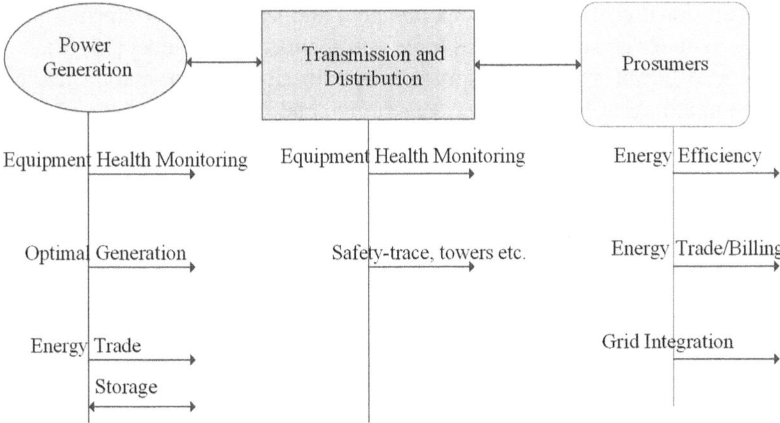

FIGURE 5.5 Benefits of IoT in the Value Chain of the Power Sector.

In summary, some of the key uses of IoT include, but are not limited to, the following:

- Peer-to-peer electricity trade,
- Fault detection for proactive maintenance activities,
- Optimization of energy storage systems,
- Optimal utilization of generation assets,
- Value-add consumer services.

This is shown in Figure 5.5.

Ancillary services are services purchased by power system operators to ensure continuous and reliable power system operation. They are grouped into system management services (ensures secure and efficient operation and monitoring of the electricity system), frequency control services (guarantees the match between generation and demand), voltage control services, and system restoration services (used for grid restoration after major outages).

Intermittent renewable energy sources are still capable of taking part in the ancillary service markets when allowed to do so. Wind and solar generators, e.g., are allowed to provide balancing services in many jurisdictions. They are increasingly being made responsible for their own imbalance in the grid. They are being penalized for incorrect forecasts; thus, deployment of IoT and advanced weather forecast technologies has never been more critical [51–56].

5.4 IMPACT OF INDUSTRY 4.0 ON POWER SYSTEM CONTROL AND OPTIMIZATION

The investment in smart grids with associated 4IR technologies will definitely result in improved efficiency and reliability of the electricity supply, integration of more renewable energy into the grid, quicker power restoration after system disturbances, lower operational cost, reduced peak demand, and less carbon footprint, ultimately resulting in better power system resource optimization and control [57–61].

Advanced power system transmission and distribution automation has enabled real-time handling of load changes, generation, and failure conditions of the power transmission and distribution system without human intervention. This necessitates automated control of devices such as switches, telemetry, relaying, and processing of critical information to the utility national and regional control centers.

Deeper penetration of intermittent renewable energy generation technologies will require progressive technologies for monitoring and optimal utilization of the generated energy. Power utilities can engage providers to aggregate their resources and offer grid services via virtual power plants and virtual transmission lines to help enhance the overall stability of the power system due to the intermittent nature of renewables, as well as assisting in the deferral of expensive system upgrades and expansion. The deployment of advanced and robust data analytics is critical to the successful deployment of renewable energy aggregation grid services.

5.5 OPPORTUNITIES AND CHALLENGES FOR POWER SYSTEM OPERATORS AND REGULATORS IN THE WAKE OF INDUSTRY 4.0

Power utilities and companies continuously collect huge amounts of data on a continuous basis through billions of sensors, telemetering, thermostats, grid equipment, wireless transmission, network communication, cloud computing, drones, and robots, among other ways. These data show the real-time state of machines and equipment, energy dispatch, load monitoring and control, weather patterns, transmission and distribution system operation and control, and power quality, among other forms of data. This is good for better monitoring and control of power generation, power network operation and customer control, and resource optimization [58, 60].

Data utilization in an efficient, enterprising, and sustainable way to improve operational efficiencies while driving down operational costs and improving their bottom line in the wake of a disrupted business model is a noble investment for power utilities.

Through big data enterprise analytics, power utilities can make the best use of available power generation, transmission, and distribution assets. This should ultimately result in an improvement in the energy production efficiency and lowering of the production, transmission, and distribution costs. Data analysis is also critical in peak demand management and demand-side management.

A multiplicity of sensors installed in a power system network and customer equipment continuously collects very sensitive data. Keeping those data secure is very important not only for consumer trust but also to prevent costly system operational mistakes. There is a need to formulate progressive laws and regulations that define and regulate ownership and the use of large amounts of data being collected and collated.

The ever-present and rapidly evolving cybersecurity threats need to be considered when making the initial decision to invest in AI techniques. As more devices connect to the IoT, companies will face a greater challenge of fragmentation, interoperability, and data security. This includes higher risks such as taking over or hacking and shutting down power plants and switching off entire power grids.

If not well managed, 4IR has a big risk of heralding post-work era with bounded freedom, extreme nationalism, bounded democracy, and vague ethics. The energy generation, transmission, and distribution industry is bound to encounter the same challenges, thus the need for progressive legal and regulatory frameworks across regional and national utilities, states, countries, and regional power pools.

The electric power sector is highly regulated; as a matter of fact, policymakers will play a crucial role in the faster adoption of new and emerging technologies. The regulation is largely informed by the fact that energy is a basic human need (which attracts a lot of government interest), and the pressure on utilities to make profit for shareholders. Policymakers should make it possible for utilities and other players to set up small-scale demonstration and pilot projects, e.g., by creating regulatory sandboxes that loosen the regulations of the electric power sector to allow for experimentation. Only decisive action by utilities can direct the transformation of electric power systems and deliver reliable energy more cheaply, cleaner, and efficiently in addition to an improved return on investment.

The current system of public policy, laws, and regulatory frameworks was developed during the Second and Third Industrial Revolutions. During the said revolutions, decision-makers had time to analyze the issues at hand and develop the necessary regulatory framework. The fast pace of change brought about by the 4IR has made regulatory work a big and fairly fluid challenge. Embracing agile governance coupled with continuous reinvention in the face of rapid disruption and innovation will be the way to go for utilities that wish to remain afloat.

Industry regulators and law makers need to come together in formulating progressive regulations and laws in the face of emerging technological advancement. Applicable benefits and penalties must also be formulated. There is a need to identify and reward companies that invest in optimized and integrated smart grids that are not cheap.

Transformation to second-generation Utilities 2.0 or Energy 4.0 that can accommodate and thrive alongside distributed renewables, energy storage, and advanced energy management will be very important for business growth and sustainability of existing power utilities. This will enable the creation of a market place for utilities and non-utility participants to provide their services as the energy democracy takes root. Industry captains need to set up a fair and progressive regulatory environment for this to take root.

Progressive tariff structures such as granular time-of-use rates should encourage planning and investment in intermittent renewables that more closely match grid requirements, as well as driving customer behavioral change.

There is a need to develop an innovative regulatory framework that will produce complex business models that provide incentives for cost efficiency, reliability, grid resilience, carbon emission, customer centricity, and performance as opposed to capital expenditure alone. Cost-of-service regulatory frameworks must be replaced with systems that allow utilities to earn incentives for exceeding expectations, along with penalties for not meeting them.

The complete optimization of the 4IR potential in power systems will also require a paradigm change in the engineering training. Human knowledge has evolved from knowledge formulation during the First Industrial Revolution followed by knowledge evolution and knowledge distribution for the Second and the Third Industrial Revolution, respectively. 4IR has brought on board the era of rapid knowledge mutation.

The challenging elements for 21st-century engineering education and training include, but are not limited to, the following [36]:

- The impact of globalization and digitization on all fields of human life and in all areas of society;
- The enormous and driven growth of the area of engineering;
- Terrific acceleration of the life cycles of engineering products;
- The changing focus of engineering shifts from more technical subjects to subjects directed to information technologies and day-to-day life;
- The increasing complexity of technical subjects that are more and more connected to nontechnical subjects;
- The requirements of a sustainable and circular economy.

The engineering education and training ecosystem must be dramatically renewed in a disruptive manner.

- Instead of acquiring new knowledge, we must teach and train on new and relevant competencies and skills and grow individual identities.
- Instead of classroom-based teaching, we have to ensure a context-aware continuous and personalized learning and training.
- Instead of life-long degrees and diplomas, we should have more and more of the on-demand and in-context continuous accreditation of qualifications and professional development.
- Fundamental education in science subjects.
- Inquiry-based, engineering education-based, and problem-based projects.
- Teach the students how to learn as opposed to just getting educated.
- A new engineering learning pedagogy based on shortening learning phases, active learning, game-based learning, project-based learning, and inquiry learning space.
- Doing is better than thinking.
- Provide an environment that encourages and facilitates practical and experiential learning.
- Integrative learning—deep level connection between the process of learning, reflective self-awareness, and experiential learning.

As a result of 4IR-inspired training and education, several spheres of intelligence, namely, emotional intelligence, contextual intelligence, ethical intelligence, inspired intelligence, socratic intelligence, strategic intelligence, physical intelligence, entrepreneurial intelligence, and transdisciplinary intelligence, will be heavily disrupted, resulting in more progressive engineering training and practice.

Virtual, augmented, and mixed reality are expected to improve manpower training proficiency, thus resulting in better quality of maintenance activities and enhanced equipment and machinery operational safety. Better, more interactive, and contextualized presentation of technical designs, drawings, and diagrams will go a long way in the improvement of the training and power system design and operations. This will assist in improving the conventional 2D single-line diagrams used in many power system drawings as power systems become more and more interconnected and complex.

The application of digital technologies that result in transformative change involves many challenges in the energy sector and in all other sectors. A qualified and skilled workforce is the first and foremost need from a managerial perspective. All employees need digital knowledge and skills at different levels, regardless of their organizational roles. A clear managerial vision and a well-defined digital strategy are other vital requirements for an organization to take steps toward the digital transformation.

Transformation of organizations, processes, and technologies is driven by digital transformation. Such reforms are often faced with resistance at various levels of management. Change management, which is mainly focused on overcoming employee resistance, is a critical managerial skill for businesses that cannot expand without ongoing transformation.

Legacy systems and unreliable data quality are some of the technological barriers to digital transformation in the power industry. Many energy companies are still using legacy systems that are not compatible with emerging technologies, which can slow the adoption of new systems. The quality and consistency of data are also critical for digital transformation, but most energy companies struggle with poor quality and not-as-useful data.

Despite the advantages offered by smart grids and grid-edge technologies, concerns related to privacy, safety, and loss of control have emerged in society. Therefore, the acceptance of consumers has become a critical factor in the digital transformation of the electricity sector. To increase acceptance, it is important to raise awareness, educate consumers, and highlight the benefits of smart meters, such as providing precise billing, reducing energy consumption, and lowering costs by adjusting usage.

The implementation of technologies in business processes is only a small part of the digital transformation. Technologies must generate additional value and synergy for customers, the business, and other key stakeholders. To succeed in digital transformation, utilities must focus on two complementary activities: reshaping customer value propositions and transforming their operations using digital technologies for greater customer interaction and collaboration. This is where the overall transformation to Society 5.0 becomes very important.

Proper digital transformation of an organization requires that the technological advances, peoples, and the business models evolve in a synchronized manner. People should always be at the center of any sustainable progress, as illustrated in Figure 5.6.

Transformation to Society 5.0 will play a critical role in the success of the digital technologies deployed in the energy sector. This is both for the professionals and for the work force in organizations. People need to gradually incorporate digital technologies into every aspect of their daily lives. This affects their adoption of digital trends, identity, and privacy concepts, as well as their employment, communication style, and way of life. The net result is an improvement in the quality of life, better output, and more innovations at the workplace. This will enable organizations

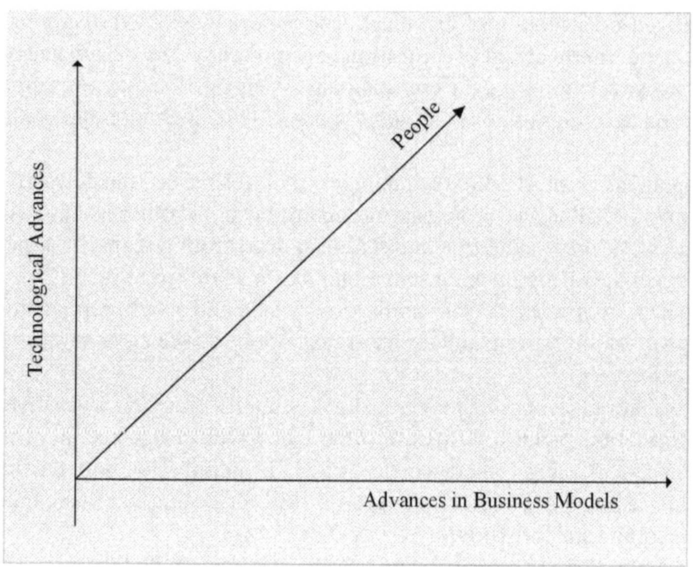

FIGURE 5.6 Confluence of Society–Technology and Business Model Transformation.

to cope better with emerging customer demands and, consequently, to survive in the future. It also enables businesses to compete more successfully in a continually changing economy due to technological advances. Effective digital transformation management provides organizations with operational and productive advantages.

5.6 SUMMARY

This chapter started by looking at the transformation from Industry 1.0 to Industry 4.0. The latter is a confluence between the physical, digital, and biological space. Good examples are AI, IoT, machine learning, cloud computing, robotics, additive manufacturing, digital twins, nanotechnology, bioinformatics, and genetic engineering, all tending to converge into one.

After that, the chapter looked at the impact of diverse Industry 4.0 innovations such as drones, data analytics robots, IoT, and digital twins on the power system value chain from generation to retail space. This includes enabling smooth integration of the variable renewable energy mix into the grid.

REFERENCES

[1] H. Okamoto, "Utility 3.0: Japan's utility of the future," *Electra*, vol. 311, 2020.

[2] A. Mcafee and E. Brynjolfsson, "Machine, platform, crowd: Harnessing the digital revolution," in *Harnessing Our Digital Future*, W. W. Norton & Company, 2017.

[3] L. Cozzi and V. Franza, *Digitalisation: A New Era in Energy, Extract from Digitalisation & Energy Report*, IEA, 2017.

[4] P. Maroufkhani et al., *Digital Transformation in the Resource and Energy Sector: A Systematic Review*, Resource Policy, 2022.

[5] U. Cali et al., *Digitalisation of Power Markets and Systems Using Energy Informatics*, Springer, 2021.
[6] K. Osmundsen, "Competences for digital transformation: Insights from the Norwegian energy sector," in *Proceedings of the 53rd Hawaii International Conference on System Sciences, HICSS, Wailea, Hawaii*, pp. 4326–4335, 2020.
[7] N. Verina and J. Titko, "Digital transformation: Conceptual framework," in *Proceedings of the International Scientific Conference on Contemporary Issues in Business, Management, and Economics Engineering, Vilnius, Lithuania*, pp. 9–10, 2019.
[8] F. Zaoui and N. Souissi, "Roadmap for digital transformation: A review of the literature," in *Proceedings of the 7th International Conference on Emerging Inter-Networks, Communication and Mobility (EICM), Leuven, Belgium*, vol. 175, August 2020.
[9] E. Piccinini et al., "Changes in the producer-consumer relationship towards digital transformation," in *Proceedings of the 12th International Conference on Wirschaftsinformatik, Osnabruck, Germany*, pp. 1634–1648, 2015.
[10] J. Hustoft and B. Weber, *The Impact of Digital Transformation on the Electric Power Industry, An Explorative Study of the Largest Norwegian Distribution System Operators*, Norwegian School of Economics, 2019.
[11] L. Cozzi and V. Franza, *Digitalisation: A New Era in Energy*, Digitalisation & Energy, IEA, 2017.
[12] R. Trzaska et al., "Digitalisation business strategies in energy sector: Solving problems with uncertainty under industry 4.0 conditions," in *Energies*, vol. 14(23), pp. 1–21, 2021.
[13] F.E. Abrahamsen et al., "Communication technologies for smart grid: A comprehensive survey," in *Sensors (Basel)*, vol. 21(23), 2021.
[14] T. Ackermann et al., "Distributed generation: A definition," in *Electric Power Systems Research*, vol. 57(3), pp. 195–204, 2001.
[15] T. Adefarati and R.C. Bansal, "Integration of renewable distributed generators into the distribution system: A review," in *IET Renewable Power Generation*, vol. 10(7), pp. 873–884, 2016.
[16] A. Alimhan et al., "The fourth industrial revolution: Toward energy 4.0 in Kazakhstan," in *International Conference on Advanced Communication Technology*, pp. 527–532, 2019.
[17] J.H. Ang, "Energy-efficient through-life smart design, manufacturing, and operation of ships in an industry 4.0 environment," in *Energies*, vol. 10(5), 2017.
[18] M.A. Berawi, "The role of industry 4.0 in achieving the sustainable development goals," in *International Journal of Technology*, vol. 10, pp. 644–647, 2019.
[19] F. Shrouf et al., "Smart factories in industry 4.0: A review of the concept and energy management approached in production based on the internet of things paradigm," in *Proceedings of the IEEE International Conference on Industrial Engineering and Engineering Management, Malaysia*, pp. 697–701, 2014.
[20] S. Lange et al., "Digitalization and energy consumption: Does ICT reduce energy demand," in *Ecological Economics Journal*, vol. 176(42), pp. 1–29, 2020.
[21] G. Ghobakhloo and M. Fathi, "Industry 4.0 and opportunities for energy sustainability," in *Journal of Clean Production*, vol. 252, pp. 210–220, 2021.
[22] P.F. Borowski, "Digitization, digital twins, blockchain, and industry 4.0 as elements of the management process in enterprises in the energy sector," in *Energies*, vol. 14(7), pp. 1–20, 2021.
[23] D. Rangel-Martinez et al., "Machine learning on sustainable energy: A review and outlook on renewable energy systems, catalysis, smart grid, and energy storage," in *Chemical Engineering Research and Design Journal*, pp. 414–441, 2021.
[24] M.B. Mollah et al., "Blockchain for future smart grid: A comprehensive survey," in *IEEE Internet Things Journal*, pp. 18–43, 2020.

[25] Z. Alavikia and M. Shabro, "A comprehensive layered approach for implementing Internet-of-things-enabled smart grid: A survey," in *Digital Communications Journal*, pp. 388–410, 2022.

[26] Y. Guo et al., "When blockchain meets smart grids: A comprehensive survey," in *High-Confidence Computing Journal*, vol. 2(2), pp. 1–25, 2022.

[27] S. Aman et al., "Energy management systems: State-of-the-art and emerging trends," in *IEEE Communications Magazine*, pp. 114–119, 2013.

[28] M. Ghobakhloo, "Industry 4.0, Digitization, and opportunities for sustainability," in *Clean Production Journal*, vol. 252(10), pp. 210–215, 2020.

[29] K. Lee, "The internet of things (IoT): Applications, investments, and challenges for enterprises," in *Business Horizons Journal*, pp. 431–440, 2015.

[30] K. Mekki et al., "A comparative study of LPWAN technologies for large-scale IoT deployment," in *ICT Express*, pp. 1–7, 2019.

[31] A.F.S. Borges et al., "The strategic use of artificial intelligence in the digital era: Systematic review of the literature and future research directions," in *International Journal of Information Management*, vol. 57, pp. 1–16, 2020.

[32] M. Andoni et al., "Blockchain technology in the energy sector: A systematic review of challenges and opportunities," in *Renewable and Sustainable Energy Reviews*, pp. 143–174, 2019.

[33] U. Bodkhe et al., "Blockchain for industry 4.0: A comprehensive review," in *IEEE Access*, pp. 79764–79800, 2020.

[34] W.A. Rezazade et al., "Debating big data: A review of the literature on realizing value from big data," in *Journal of Strategic Information Systems*, pp. 191–209, 2017.

[35] D.A. Domalewska, "Longitudinal analysis of the creation of environmental identity and attitudes toward energy sustainability using the framework of identity theory and big data analysis," in *Energies*, vol. 14, 2021.

[36] X. Han et al., "Augmented reality in professional training: A review of the literature from 2001 to 2020," in *Applied Sciences Journal*, vol. 12(3), pp. 12–34, 2022.

[37] B.L. Ludlow, "Virtual reality: Emerging applications and future directions," in *Rural Special Education*, pp. 3–10, 2015.

[38] Y. Zheng et al., "An application framework of digital twin and its case study," in *Journal of Ambient Intelligence and Humanized Computing*, pp. 1141–1153, 2019.

[39] A. Fuller et al., "Digital twin: Enabling technologies, challenges, and open research," in *IEEE Explore*, vol. 8, 2021.

[40] N. Komninos et al., "Survey on smart grid and smart home security: Issues, challenges, and countermeasures," in *IEEE Communications Survey and Tutorials*, pp. 1933–1954, 2014.

[41] Z. Fan et al., "Smart grid communications: Overview of research challenges, solutions, and standardisation activities," in *IEEE Communications Survey and Tutorials*, pp. 21–38, 2013.

[42] R. Singh et al., "Internet of things based on home automation for intrusion detection, smoke detection, smart appliance, and lighting control," in *International Journal of Scientific and Technology Research*, pp. 3702–3707, 2019.

[43] J. Moradi et al., "Internet of energy (IoE) in smart power systems," in *Proceedings of the 2019 IEEE 5th Conference on Knowledge Based Engineering and Innovation, Tehran*, pp. 627–636, 2019.

[44] B. Celik et al., "Management of electrical energy in residential areas through coordination of multiple smart homes," in *Renewable and Sustainable Energy Reviews*, pp. 260–275, 2017.

[45] F. Al-Turjman and M. Abujubbeh, "IoT-enabled smart grid via SM: An overview," in *Future Generation Computer Systems Journal*, pp. 579–590, 2019.

[46] L. Chettri and R.A. Bera, "Comprehensive survey on the internet of things (IoT) toward 5G wireless systems," in *IEEE Internet Things Journal*, pp. 16–32, 2020.

[47] K. Siozios et al., *IoT for Smart Grids*, Springer, 2019.

[48] M. De Rosa et al., "Flexibility assessment of a combined heat-power system (CHP) with energy storage under a real-time energy price market framework," in *Thermal Science and Engineering Progress*, pp. 426–438, 2018.

[49] G. Zhang et al., "Online energy management for microgrids with CHP co-generation and energy storage," in *IEEE Transactions on Control Systems Technology*, pp. 533–541, 2020.

[50] M. Pau et al., "A cloud-based smart metering infrastructure for distribution and automation of grid services," in *Sustainable Energy Grids and Networks*, pp. 14–25, 2018.

[51] M. Medojevic et al., "Energy management in industry 4.0 ecosystem: A review of possibilities and concerns," in *DAAAM International Symposium on Intelligent Manufacturing and Automation*, pp. 674–680, 2018.

[52] D. Campo et al., "IoT solution for energy optimisation in industry 4.0: Issues of real-life implementation," in *Proceedings of the 2018 Global Internet of Things Summit, Bilbao*, pp. 1–6, 2018.

[53] P. Solic et al., "Internet of things (IoT): Opportunities, issues and challenges towards a smart and sustainable future," in *Journal of Clean Production*, vol. 274, pp. 121–133, 2020.

[54] B. Zohuri, "Hybrid energy systems," in *Hybrid Renewable Energy Systems*, Springer, pp. 1–38, 2018.

[55] E.T. Hale et al., *Potential Roles for Demand Response in High-Growth Electric Systems with Increasing Shares of Renewable Generation*, National Renewable Energy Lab (NREL), 2018.

[56] F. Un-Noor et al., "A comprehensive study of key components, technologies, challenges, impacts, and future direction of development for electric vehicles," in *Energies*, vol. 10, pp. 1–84, 2017.

[57] M.L. Tuballa and M.L. Abundo, "A review of the development of smart grid technologies," in *Renewable and Sustainable Energy Reviews*, pp. 710–725, 2016.

[58] A. Salam et al., "The future of emerging IoT paradigms: Architectures and technologies," in *Preprints*, p. 1803020, 2019.

[59] A. Gabash, "Framework for optimal real-time power flow under wind energy penetration," in *Energies*, vol. 10(4), pp. 1–11, 2017.

[60] M. Sanduleac et al., "Next-generation real-time smart metres for ICT based assessment of grid data inconsistencies," in *Energies*, vol. 10(7), pp. 12–30, 2017.

[61] U. Damisa et al., "Towards blockchain-based energy trading: A smart contract implementation of energy double auction and spinning reserve trading," in *Energies*, vol. 15, 2022.

6 Trends and the Future of Power Systems

6.1 SOCIETY 5.0 AND POWER SYSTEMS

In January 2016, Japan postulated the concept of Society 5.0 characterized by the seamless integration among physical, digital, and cyberspace (cyber–physical integration). This is expected in result to a super-smart, data-driven [from Internet of Things (IoT) systems], humanity-centered society where there is no line between man and machine. The ultimate vision is to create the Internet of humans, beyond the IoT [1–15].

This is shown in Table 6.1.

Mental well-being remains a critical human need for better productivity. Better economic output and growth will be realized if we utilize the Internet of humans to digitize human activity and create a humanity-centered society that supports happiness and better and all-rounded well-being.

The underlying objective and vision are to provide a society that balances between technological progression, economic empowerment, and endowment while striving to solve social and societal challenges for the greater good of all members of the society. One way is by rapid processing of a vast array of data collected and analyzed to produce useful information in the real world that enables making our lives more comfortable in all spheres of life, including healthcare, shopping, energy usage,

TABLE 6.1
Evolution of Technology and Human Development

Phase of Development	Society 1.0	Society 2.0	Society 3.0	Society 4.0	Society 5.0
Key aspects	Hunter and gatherers/ capture Use of stone tools Locomotion by foot Nomadic lifestyle	Agrarian/farming Metallic tools Movement by ox/horses Fortified cities and towns	Industrial/mass production Mechanical machines Metallic and plastic tools Machine transport—cars, boats, planes, etc. Industrial cities (functional)	Information age Use of computers Semiconductors and computer chips Multi mobility Highly connected cities (profitable)	Super-smart Merger of physical and cyber aspects Materials 5.0—nanomaterials and others Autonomous vehicles Smart cities (humanity-centered)

 DOI: 10.1201/9781032665290-6

education, transport, and entertainment, by offering practical solutions. This underpins the whole concept of Society 5.0 as being people-centered [16–35].

On the other hand, Society 5.0 promises an era where all aspects of power systems are brought under full control, right from the utility day-to-day operations to the empowered prosumers, e.g., switching of appliances or gadgets by pointing at them or by blinking of an eye. In general, Society 5.0 envisages the development of local and decentralized power grids to supply clean, affordable, and sustainable power to all.

Society 5.0 is ultimately anchored on Industry 5.0 (beyond Industry 4.0) and promises to provide businesses, companies, and individuals better solutions that will guarantee optimal utilization of resources, better social cohesion, better quality jobs, and better customer personification and centricity. This will be a value-driven and people-centric society as opposed to technology-only-driven societies where there is a lack of proactive solving of human and societal problems.

One example of the use of data for societal good is sharing information on the level of environmental pollution in open-source platforms where any citizen can pick it up and take the necessary action or advocacy decisions. In power systems, availing a granular tariff structure to see the amount of clean renewable data to all in real time can help businesses take advantage of cheap clean energy.

Savings such as reduction on water and energy usage can be bigger societal goals as opposed to typical short-term financial gains, e.g., social return on investment, namely, the contribution toward reduction of global warming, which is not only more incentivizing but is also more appealing to investors who may feel like they are doing more for the greater good of everyone.

As such, the envisioned social and economic transformation will be driven by three elements, namely, data, knowledge, and information, as shown in Figure 6.1.

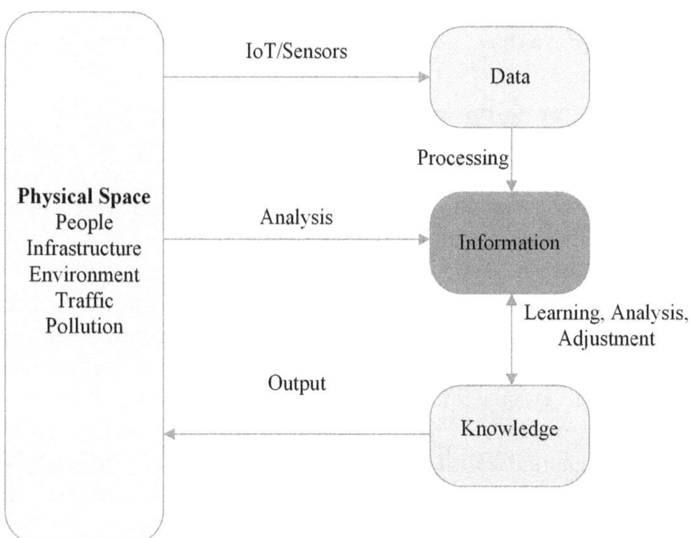

FIGURE 6.1 Physical Space–Knowledge Convergence.

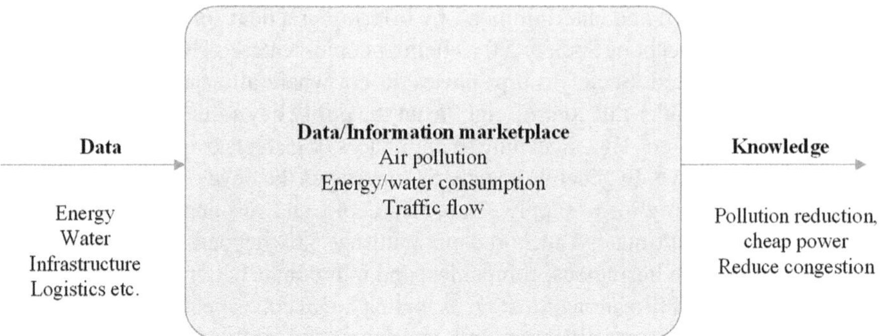

FIGURE 6.2 Creation of a Smart City Data Marketplace.

Information is data that has been analyzed and converted to meaningful and useful patterns for better, proactive, and real-time decision-making. Information is converted to knowledge when it is well analyzed, understood, and tied to best practices, patterns, and scientific laws. As such, data become useful once we convert them to information and then, more importantly, to useful knowledge. In Society 5.0, value is derived from knowledge rather than the conventional tangible assets, thus the saying that data are gold. As such, Society 5.0 envisages a knowledge-rich society that will ultimately spur more useful research and innovations anchored in the continuous processing of vast arrays of data.

We are looking at a situation where all smart cities will have a city data exchange that will serve as the marketplace for big data trading and exchange. These data include transport/traffic, energy, water, finance, pollution, and other forms of data continuously exchanged and traded in cyberspace by city dwellers and administrators. The objective is to ensure the best utilization of data for the benefit of all in business and to improve the quality of life. This is illustrated in Figure 6.2.

6.2 EVOLVING POWER SYSTEMS AND UTILITY BUSINESS MODELS

There are a number of compelling trends and compelling needs that will continue to shape the power utility of the future and beyond.

The rapid fall in renewable energy generation technologies will continue to spur their continued adoption, both at the personal investment, utility scale and the regional energy trade scale. The implication of this is the maturity of the decentralization business model and applicable power network operation technologies. A myriad of renewable energy generation technologies will ultimately further complicate the power network; however, if well managed, it will result in more reliable power systems and better energy security. This will be further fueled by price spikes, global political dynamics, and carbon footprint reduction associated with coal, oil, and natural gas [36–48].

The power sector will continue to compete for scarce resources, such as land and water, with other sectors of the economy. This will continue to fuel the need for more efficient smart grids, decentralized power systems with less emissions, efficient appliances, and more customer centricity.

The push on power utilities and the entire energy value chain to play an greater role in the reduction of carbon footprint will continue to tighten in the near future.

There will be a greater push on power utility businesses to become more resilient as the transition to modern technologies takes the center stage. This is resilience to cybersecurity threats and increases in extreme weather and climate events in the face of accelerated climate change phenomena.

Prosumers and customers expect expanded choices; better control over energy use, energy sources, and costs; and overall better customer experience.

Emerging business opportunities and disruptive technologies and trends in technological capabilities require utilities to become more innovative within the utility and to enable innovation by others. New revenue streams arising from enhanced services powered by the new grid technologies will propel forward-looking utilities toward better financial stability.

The regulators will also be expected to play a more dynamic and fluid role in the face of rapid changes in business models.

6.3 POLICY AND REGULATORY CONTROLS, CHALLENGES, AND INNOVATIONS FOR POWER SYSTEMS

More than 100 years ago, at the onset of electric utility businesses across the world, strict regulation was a necessity largely because it legitimized the utility business in the eyes of the consumer. A skeptical consumer and the general public needed assurance that electricity was a good, thus the need for the government's hand in the entire chain. This support was necessary to ensure that local governments provided the necessary support in areas such as the right of way on road corridors and other public utility spaces required for grid extension. This was also a necessity to ensure investors of stable returns from a business that has a fairly high initial outlay and high operating cost. With the assurance, each utility could focus on building its network without being undercut by competitors. Government control was also required in the establishment of grid construction and operation standards for public safety and low operational cost. This minimized the issues of conflicting customer claims, operational standards, and different offers from different power utilities. Regulated pricing guaranteed stable returns for investors pumping in huge amounts of capital required to set up the infrastructure right from generation, transmission, and distribution assets. Without this level of assurance, we would not have the huge electric industry infrastructure we have in many countries around the world today [49–52].

The road to deregulation and liberalization of the energy sector has been long and winding. The winds of change came slowly but surely, almost unnoticed. Up to around the 1980s, the most cost-competitive energy generation was done by the use of huge fossil fuels, natural gas, and coal-fired power plants. However, with the advent of better technology, the emergence of more efficient and smaller turbines and generators led to a relook at this model. Many industrial users could for the first time generate their own energy at economic prices. There was also pressure and lobbying from major consumers to be allowed to get their power supply from generators that offered the lowest prices and the best consumer services. The upheavals in global oil prices and pressure of the global warming challenge further led to the need to continuously innovate toward cleaner, possibly renewable sources of energy. This metamorphosis also sowed the seed for many governments to see the need to sell their stake in power utility companies to private players who could ordinarily do a better job. Deregulation to ease up the many stringent rules of the utility business assured investors of good returns on their investment. Thus, deregulation almost follows privatization as a prerequisite for private players to pump in money.

The push toward deregulation began with the realization of the many benefits and value-added services that come with the introduction of competition in the electric energy value chain right from generation, transmission, distribution, and retail segments. Competition pushes business owners to innovate, operate more efficiently, and at the lowest possible cost while offering the best customer experience. This was the foundation for the granular time-of-use tariff system in use in many jurisdictions in the world today.

The power utility industry has been highly regulated and controlled by various government entities for the last 100 years. The level of control includes granting utilities exclusive areas of operation in terms of energy transmission and distribution, setting out certain minimum obligations to customers and the general public, setting out parameters for the least cost of operation model that guarantees investor returns, and prescribing business models on aspects such as grid code, among other legal and regulatory controls. With the continued penetration of intermittent renewable energy into the grid, the rise of consumers and evolving business models in areas such as energy storage, the need for a review and relook of the existing regulatory framework has been long overdue.

6.4 SUMMARY

This chapter began by looking at the concept of the envisaged Society 5.0 that promises to have no boundary between man and machine. This is the next level Industry 4.0 that is characterized by the confluence between physical, biological, and digital spaces. The result will be a super-smart, humanity-driven society geared toward solving of social and societal challenges.

There are a number of compelling trends and compelling needs that will continue to shape the power utility of the future and beyond. Society 5.0 will bring all elements of power systems under full control, characterized by the maturity of the decentralized business model and applicable power network operation technologies.

The industry regulators will be expected to play a more critical role in the ever-fluid business environment.

REFERENCES

[1] C. Abbey et al., "Powering through the storm: Microgrid operation for more efficient disaster recovery," in *IEEE Power and Energy Magazine*, vol. 123, pp. 67–76, 2014.

[2] S. Aggarwal and E. Burgess, "Performance-based models to address utility challenges," in *The Electricity Journal*, vol. 27(6), pp. 48–60, 2014.

[3] M. Bazilian et al., "Accelerating the global transformation to 21st century power systems," in *The Electricity Journal*, vol. 26, pp. 39–51, 2013.
[4] Y. Chen et al., "Evidence on the impact of sustained recovery exposure to air pollution on life expectancy from China's Huai River policy," in *Proceedings of the National Academy of Sciences*, vol. 110, pp. 12936–12941, 2013.
[5] Citi, *Energy Darwinism: The Evolution of the Energy Industry*, Global Perspectives & Solutions Series, Cambridge University Press, 2013.
[6] J. Cochran et al., *Market Evolution: Wholesale Electricity Market Design for 21st Century Power Systems*, NREL/TP-6A20-57477, National Renewable Energy Laboratory, 2013.
[7] Commonwealth Scientific and Industrial Research Organisation (CSIRO), *Change and Choice: The Future Grid Forum Analysis of Australia's Potential Electricity Pathways to 2050*, CSIRO, 2013.
[8] Electric Power Research Institute (EPRI), *The Integrated Grid: Realising the Full Value of Central and Distributed Energy Resources*, EPRI, 2014.
[9] Environmental Protection Agency (EPA), *The Benefits and Costs of the Clean Air Act from 1990 to 2020*, EPA, 2011.
[10] C. Gao and Y. Li, "Evolution of China's power dispatch principle and the new energy: Saving power dispatch policy," in *The Energy Policy*, vol. 38(11), pp. 7346–7357, 2010.
[11] E. Gaudchau et al., "Business models for renewable energy, energy-based mini-grids in non-electrified regions," in *Proceedings of the 28th European PV Solar Congress Energy Conference*, vol. 4, 2013.
[12] K.N. Gratwick and A. Eberhard, "Demise of the standard model for power sector reform and the emergence of hybrid power markets," in *The Energy Policy*, vol. 36(10), pp. 3948–3960, 2008.
[13] C. Greacen et al., *A Guidebook on Grid Interconnection and Islanded Operation of Mini-Grid Power Systems Up to 200 Kw*, LBNL Report, 2013.
[14] R. Green II et al., "The impact of plug-in hybrid electric vehicles on the environment, distribution networks: A review and outlook," in *Renewable and Sustainable Energy Reviews*, vol. 15(1), pp. 544–553, 2011.
[15] M. Hogan, *Power Markets: Aligning Power Markets to Deliver Value*, America's Power Plan, 2013.
[16] IEA Renewable Energy Technology Deployment (IEA-RETD), *Residential Prosumers Drivers and Policy Options (RE-Prosumers)*, IEA-RETD, September 2019.
[17] ISGAN, *Flexible Power Delivery Systems: An Overview of Policies and Regulations, Expansion Planning and Market Analysis for the United States and Europe*, ISGAN Discussion Paper, Annex 6: Power T&D Systems, 2013.
[18] T. Fujii et al., "Consideration of the service strategy of Japanese electric manufacturers to realise super smart society (SOCIETY 5.0)," in *Proceedings of the Programmieren für Ingenieure und Naturwissenschaftler*, Springer Science and Business Media LLC, pp. 634–645, 2018.
[19] M.A. Nagahara, "Research project of society 5.0," in *Proceedings of the 2019 IEEE Conference on Control Technology and Applications (CCTA), Kitakyushu, Japan*, pp. 803–804, 2019.
[20] V. Potocan et al., "Society 5.0: Balancing of industry 4.0, economic advancement, and social problems," in *Kybernetes*, pp. 794–811, 2020.
[21] S. Takakuwa et al., "Industry 4.0 in Europe and East Asia," in *DAAAM Conference, Vienna*, pp. 0061–0069, 2018.
[22] M.A. Berawi, "Managing nature 5.0 in industrial revolution 4.0 and society 5.0 era," in *International Journal of Technology*, vol. 10(2), pp. 222–225, 2019.
[23] L. Melnyk et al., "The effect of industrial revolutions on the transformation of social and economic systems," in *Problems and Perspectives in Management*, pp. 381–391, 2019.

[24] H.I. Elim and G. Zhai, "Control system of multitasking interactions between society 5.0 and industry 5.0: A conceptual introduction & its applications," in *Journal of Physics: Conference Series*, vol. 1463, pp. 1–9, 2020.
[25] M.E. Gladden, "Who will be the members of society 5.0? Toward an anthropology of technologically posthumanized future societies," in *Social Science Journal*, vol. 8(5), pp. 1–39, 2019.
[26] B. Aquilani et al., "The role of open innovation and value co-creation in the challenging transition from industry 4.0 to society 5.0 toward a theoretical framework," in *Sustainability Journal*, vol. 12, 2020.
[27] C. Serpa and S. Ferreira, "Society 5.0 and social development: Contributions to a discussion," in *Journal of Organizational Management Studies*, pp. 26–31, 2018.
[28] I. Gustiana et al., "Society 5.0: Optimisation of the sociotechnical system in poverty reduction," in *IOP Conference Series: Materials, Science and Engineering*, vol. 662, 2019.
[29] M. Kinoshita, *Japan on the New Industrial Revolution (NIR): Direction and Its Global Implications for Inclusive and Sustainable Industrial Development*, University of Tokyo, 2019.
[30] K. Fukuda, "Science, technology and innovation ecosystem transformation toward society 5.0," in *International Journal of Production Economics*, vol. 220, pp. 1–14, 2020.
[31] A.V. Mavrodieva and R. Shaw, "Disaster and climate change issues in Japan's society 5.0—a discussion," in *Sustainability Journal*, vol. 12(6), pp. 1–17, 2020.
[32] M. Fukuyama, "Society 5.0: Aiming for a new human-centred society in Japan," in *Spotlight*, pp. 47–50, 2018.
[33] Y. Shiroishi, "Better actions for society 5.0: Using AI for evidence-based policy making that keeps humans in the loop," in *Computer (Long Beach, Calif.)*, vol. 52, 2019.
[34] S. Arsovski, "Quality of life and society 5.0," in *Proceedings on Engineering Sciences*, pp. 775–780, 2019.
[35] K. Matsuda et al., "Technologies of production with society 5.0," in *Proceedings of the 6th International Conference on Behavioural, Economic and Socio-Cultural Computing (BESC), Beijing, Jiaotong University, China*, pp. 1–4, 2019.
[36] R. Foresti et al., "Smart society and artificial intelligence: Big data scheduling and the global standard method applied to smart maintenance," in *Engineering*, pp. 835–846, 2020.
[37] A. Lavalle et al., "Improving the sustainability of smart cities through visualization techniques for big data from IoT devices," in *Sustainability*, vol. 12(14), pp. 1–17, 2020.
[38] F. Longo et al., "Value-orientated and ethical technology engineering in industry 5.0: A human-centric perspective for the design of the factory of the future," in *Applied Sciences*, vol. 10(12), pp. 1–25, 2020.
[39] R. Carraz and Y. Harayama, "Japan's innovation systems at the crossroads: Society 5.0 in panorama: Insights into Asian and European affairs," in *Digital Asia*, Panorama Publishing House, vol. 3, pp. 33–45, 2018.
[40] H. Sedjelmaci, "Attack detection approach based on a reinforcement learning process to secure 5G wireless network," in *Proceedings of the IEEE International Conference on Communications Workshops (ICC Workshops), Dublin, Ireland*, pp. 1–6, 2020.
[41] M. Vaga et al., "Techniques for secure automated operation with Cobots participation," in *Proceedings of the 21st International Carpathian Control Conference (ICCC), High Tatras, Slovakia*, pp. 1–4, 2020.
[42] S. Duangsuwan et al., "Development on air pollution detection sensors based on NB-IoT network for smart cities," in *Proceedings of the 18th International Symposium on Communications and Information Technologies (ISCIT), Bangkok, Thailand*, pp. 313–317, 2018.

[43] M. Shibata et al., "Toward an efficient search method to capture the future MOT curriculum based on the society 5.0," in *Proceedings of the 2017 Portland International Conference on Management of Engineering and Technology (PICMET), Portland*, pp. 1–7, 2017.
[44] J. Alvarez-Cedillo, "Actions to be taken in Mexico towards education 4.0 and society 5.0," in *International Journal or Evaluation and Research in Education*, vol. 8(4), pp. 693–698, 2019.
[45] A. Abbasi and M.M. Kamal, "Adopting industry 4.0 technologies in citizens' electronic engagement considering sustainability development," in *Information Systems*, Springer International Publishing, vol. 381, pp. 304–313, 2020.
[46] B. Salgues, "Societies 5.0 and the management of the future," in *Society 5.0*, John Wiley & Sons, Inc., pp. 91–119, 2018.
[47] Y. Zengin et al., "An investigation of industry 4.0 and society 5.0 in the context of sustainable development goals," in *Sustainability*, vol. 13(5), pp. 1–16, 2021.
[48] C.A.T. Romero et al., "Synergy between circular economy and industry 4.0: A review of the literature," in *Sustainability*, vol. 13(8), pp. 1–18, 2021.
[49] B. Salgues, "From society 5.0 to its associated policies," in *Society 5.0*, John Wiley & Sons, Inc., pp. 23–36, 2018.
[50] S. Serpa and C.M. Ferreira, "Society 5.0 and sustainability digital innovations: A social process," in *Journal of Organizational Culture Communications and Conflict*, pp. 1–14, 2019.
[51] C.A. Tavera et al., "Software architecture for planning educational scenarios by applying an agile methodology," in *International Journal of Emerging Technologies in Learning*, pp. 132–144, 2021.
[52] C.A. Tavera et al., "Wearable wireless body area networks for medical applications," in *Computational and Mathematical Methods in Medicine*, pp. 1–9, 2021.

Index

For Product Safety Concerns and Information please contact our EU representative GPSR@taylorandfrancis.com Taylor & Francis Verlag GmbH, Kaufingerstraße 24, 80331 München, Germany

Batch number: 10397790

Printed by Printforce, the Netherlands